Chemistry for Energy

Chemistry for Energy

M. Tomlinson, EDITOR

Chemical Institute of Canada

ASSOCIATE EDITORS

T. E. Rummery
D. F. Torgerson
A. G. Wikjord

Based on a symposium

sponsored by the Chemical

Institute of Canada at the

Annual CIC Conference,

Winnipeg, June 5–7, 1978.

A C S S Y M P O S I U M S E R I E S **90**

AMERICAN CHEMICAL SOCIETY
WASHINGTON, D. C.   1979

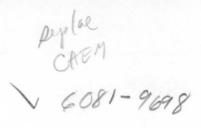

Library of Congress CIP Data

Chemistry for energy.
(ACS symposium series; 90 ISSN 0097-6156)

Includes bibliographies and index.

1. Fuels—Congresses. 2. Renewable energy sources
—Congresses. 3. Electric power production from chemical action—Congresses. 4. Atomic power-plants—Congresses.
I. Tomlinson, M. II. Chemical Institute of Canada.
III. Series: American Chemical Society. ACS symposium series; 90.

TP315.C43   662'.6   79-26175
ISBN 0-8412-0469-1   ASCMC 8   90 1–351 1979

ACS Symposium Series

Robert F. Gould, *Editor*

FOREWORD

The ACS Symposium Series was founded in 1974 to provide a medium for publishing symposia quickly in book form. The format of the Series parallels that of the continuing Advances in Chemistry Series except that in order to save time the papers are not typeset but are reproduced as they are submitted by the authors in camera-ready form. Papers are reviewed under the supervision of the Editors with the assistance of the Series Advisory Board and are selected to maintain the integrity of the symposia; however, verbatim reproductions of previously published papers are not accepted. Both reviews and reports of research are acceptable since symposia may embrace both types of presentation.

CONTENTS

ELECTRICITY PRODUCTION AND STORAGE

PREFACE

The main purpose of the Chemistry for Energy Symposium was to identify where advances in chemical knowledge and understanding are required for the development and diversification of energy sources. The contributors were invited to review the chemistry and chemical engineering aspects of various sectors of energy production from Canadian sources and to indicate important areas for research and development. With a view to encouraging participation by University chemists and engineers in energy research and development, authors were asked to select areas where expansion of fundamental chemical knowledge and basic chemical data are needed.

The symposium was divided into three sectors: fossil fuels; perpetual and renewable sources; and electricity production and storage. This publication contains a selection of papers that were presented at the symposium.

The symposium provided ample evidence that, while Canada is very well endowed with energy resources of various kinds, it will require much effort by chemists and others to make them available in useful form and at moderate price. Even with the best possible efforts and stringent conservation measures, we are unlikely to avoid shortages and importation of large amounts of oil in the 1980s. Forms of energy and energy conversions which are not economic at present then may begin to compete effectively.

Energy is distributed in several quite distinct and not readily interchangeable forms. Electricity can perform work with near 100% efficiency (it is essentially pure Gibbs free energy) but cannot be efficiently and economically stored. Fuel oil is a very compact and convenient energy store which is readily released as heat (enthalpy) that can be only partly converted to work, where so required, in some form of heat engine. Fuel gas has some characteristics of both. Thus liquid fuel and electricity have complementary functions and both are required. The distinction was occasionally overlooked by some speakers. The coming energy crisis is a deficit of liquid fuel. Supplies of cheap electricity can be maintained but can only be substituted for liquid fuel to a limited extent.

Chemistry research can help a great deal to minimize the deficit in energy supplies and the rise in energy costs and to facilitate and shift to new sources. There are three main branches of energy technology:

energy collection and concentration, conversion and storage, and application and waste-product control. The R&D emphasis is concentrated in the first two areas for novel energy sources and in the latter two for established energy sources. Fluidification is a dominant theme—the conversion of various raw energy sources, new or old, to liquid fuels.

Some papers are quite explicit as to the new chemical knowledge required for progress; in others, the requirements are implicit. Many of the chemical topics are specific to particular energy sources and to particular aspects of the technology. However, certain common threads and general requirements can be perceived, which are noted below, followed by brief discussion of some of the more specific aspects from the respective sectors of the symposium.

Surface and interface chemistry is the key to progress in many areas of energy technology. Recovery of oil from the Alberta tar sands involves separating oil–water–sand with an interfacial area of approximately $1 \, m^2/g$. Furthermore, several authors noted the major wastes problem of separating and stabilizing the solids from the residual sludge. Adhesion of fly ash to superheater tubes, which reduces the efficiency of coal use, is a problem in surface-charge phenomena. Coal can be used to extend fuel oil supplies directly if dispersions of coal in oil can be stabilized sufficiently. Conversely, knowledge of destabilization of carbonaceous colloids could lead to improvements in the dewatering of peat so that it may be used more readily as an energy source. Other aspects of surface chemistry come into the improvement of catalysts for the liquefaction or gasification of coal. Improved knowledge of various electrode interfacial processes is important in the development of better batteries and fuel cells. A specific example is the need for improved electrocatalysis to lower the overpotential for oxygen reduction. Another form of surface chemistry, the science of membranes, cellular and synthetic, is likely to be important in developing methods for trapping solar energy in chemical-stored form. Knowledge of the adsorption of dissolved radionuclides on rock can be used to provide extra assurance of retention underground of the wastes from the use of nuclear energy.

Improved knowledge of chemistry at elevated temperatures is a general requirement since all energy use and transformation processs involve temperatures higher than normal. Requirements include thermochemical and kinetic data at the temperatures of combustion and conversion reactions of fossil fuels and at the more moderate temperatures of hot aqueous solutions. Improvements in theoretical models or empirical methods are required to allow more accurate extrapolation of knowledge and data from ordinary temperatures.

Many aspects of sulfur chemistry are of concern if we are to reverse the apparent trend to increasingly acid rain while continuing to use all forms of fossil fuel whatever their sulfur content—35% H_2S in some sour gas wells. Perhaps even more important than sulfur recovery and control of emissions is to find constructive uses or innocuous methods of disposal and to determine the long-term effects of the sulfur which is prevented from immediate dispersal in the atmosphere—currently in excess of 25,000 tonnes per day accumulating on the ground in Alberta.

Improved knowledge of the structure and transformations of certain solids is desired. Notably coal, it was said, is still very imperfectly understood, despite its long use as an energy source. Improvements here would aid development of processes for conversion to liquid fuels and for recovery of solid residues in a form suitable for metallurgical coke. For new ways to tap energy sources, development of solids with high photoelectric conversion efficiency for sunlight could provide a major breakthrough. In relation to nuclear energy, knowledge of the solid-state transformations of glass and other solids is being developed to ensure that nuclear wastes are well locked into appropriate solids until they are no longer hazardous.

Advances in knowledge of the chemistry of fermentation processes will aid the exploitation of biomass energy, e.g. a more concentrated fermentation process for the production of sugar from cellulose is required if alcohol from Canada's very extensive forests is to compete with other sources of liquid fuel.

Fossil Fuels

The fossil fuel program was highlighted by five invited papers dealing with coal conversion, oil sands, desulfurization, peat, and the federal government's R&D program. In total, there were a dozen papers pertaining to Canada's fossil fuel resources, coal (four papers) and oil sands (three papers) receiving the greatest attention. Seven of these papers appear in this volume.

N. Berkowitz (Alberta Research Council) provided a stimulating account of the potential of coal in Canada's energy future. Coal can be used directly as an industrial fuel or be converted to other combustible hydrocarbons. Berkowitz described the three different conversion techniques: gasification, liquefaction, and partial conversion techniques to produce gases, oils, and solid fuels.

M. Greenfeld described unique laboratory experiments designed to stimulate and understand the complex chemistry of in-situ coal gasification. Developed at the Alberta Research Council, the gasification simulator was heavily instrumented with calorimeters and gas chromatographs to determine the enthalpy, composition, and kinetics of formation of the product gases. Computer techniques were used to calculate mass and heat balances and to test kinetic models.

K. Belinko, M. Ternan, and B. N. Nandi of the Canada Centre for Minerals and Energy Technology (CANMET) discussed the formation of mesophases during the liquefaction of noncoking coals. R. D. Humphreys (Alberta Oil Sands Technology and Research Authority) focussed on the huge energy potential of Alberta's Athabasca, Wabasca, Cold Lake, and Peace River oil-sands deposits. A staggering 2.5 trillion barrels are locked in these deposits of sand and underlying carbonate rock. About 5% of the Athabasca deposit can be extracted by mining while 30% or more is recoverable by in-situ technology. With a $150 million budget and by collaborative action with industry and university, AOSTRA's goals are to help improve surface extraction technology and land rehabilitation, to increase the efficiency of in-situ recovery, and to develop efficient processes to convert the oil sands into higher-value materials (petroleum and minerals). Humphreys described the licensing arrangements for joint AOSTRA–industry ventures and the depth of interaction with university where some 30 projects are now being funded.

A paper contributed by J. E. Desnoyers, R. Beaudoin, G. Roux, and G. Perron described the use of microemulsions as a possible tool for the extraction of oil from tar sands. Using a technique called flow microcalorimetry recently developed at the University of Sherbrooke, these researchers studied the structure and stability of organic microphases in aqueous media. These microphases can be stabilized by surfactants and can dissolve large quantities of oil. In a similar vein, D. F. Gerson, J. E. Zajic, and M. D. Ouchi (University of Western Ontario) described the extraction of bitumen from Athabasca tar sands by a combined solvent–aqueous-surfactant system.

Desulfurization of fossil fuels was the subject of an authoritative review by J. B. Hyne (Alberta Sulphur Research Institute). This is a topic of increasing importance as Canada relies more and more on sulfur-containing fuels such as tar sands and heavy oils. Hyne reviewed the present state of the chemistry and technology for both precombustion desulfurization of natural gas and crude oils and postcombustion tail-gas clean up of coals and cokes. He clearly identified areas of possible future research such as the high temperature–high pressure chemistry pertaining to in-situ desulfurization processes.

Perpetual and Renewable Sources

BIOMASS. The potential of biomass to contribute to Canada's energy needs was discussed in papers by C. R. Phillips, D. L. Granatstein and M. A. Wheatley (University of Toronto), R. Overend (Canada Department of Energy, Mines, and Resources), and M. Wayman, J. Lora, and E. Gulbinas (University of Toronto). The most energy-efficient and least costly use of biomass is the direct burning of wood, followed by gasification and liquefaction, respectively. It was estimated that 2–3% of Ontario's liquid-fuel consumption could be replaced by wood liquefaction based on 500,000 hectares of available forest. The potential would increase to 8–10% if 1.5×10^6 hectares were available. C. R. Phillips, D. L. Granatstein, and M. A. Wheatley recommended an energy program having the following order of priority: crude-oil exploration, oil sands, possibly coal liquefaction, then wood liquefaction. The need for more R&D in the wood liquefaction areas was stressed since Canada has an abundant supply of wood.

The use of anaerobic bacterial systems for conversion of animal manure into methane gas was discussed by H. M. Lapp (University of Manitoba) who described operating characteristics and factors affecting anaerobic digestion plants. M. Wayman and M. Whiteley (University of Toronto) reported on the interaction of photosynthetic and sulfate reducing bacteria in a membrane-separated anaerobic culture. This autotrophic microbial system is capable of producing a high protein biomass in one fermenter, while the other produces a high energy biomass.

SOLAR. J. Bolton (University of Western Ontario) discussed thermodynamic and kinetic limits on photochemical conversion and storage of solar energy. He stated that 25–28% efficiency should be attainable for conversion of solar energy to electricity. Some guidelines and objectives were given for research to foster development of workable fuel and electrical-generation systems using solar energy. B. L. Funt, M. Leban, and A. Sherwood (Simon Fraser University) have constructed a 100-cm^2 CdSe photoelectrochemical cell which uses a large part of the sun's energy spectrum. They assessed factors relevant to the scaling up of their cells, with the objective of attaining 1% conversion efficiency in a large converter. F. R. Smith (Memorial University) stressed that if photoelectrolysis of aqueous solutions is to become an economical process for hydrogen and oxygen production, it will be necessary to develop semiconductor anodes having band gaps matched to the solar spectrum.

Electricity Production and Storage

The final session of the Conference was devoted to discussion of the main methods of producing and storing electrical energy (batteries and fuel cells) and to a discussion of some of the chemical problems encountered during nuclear generation of electricity.

E. J. Casey (Defense Research Establishment, Ottawa) reviewed the selection of anodes and electrolytes for high-energy density storage batteries. The present state of development of batteries by using light metal anodes in nonaqueous, molten salt and solid electrolytes was reviewed, and suggestions were made on the feasibility of novel systems.

Another prospect for efficient energy conversion is the fuel cell. The different types of fuel cells presently under study or development were reviewed by G. Bélanger of Hydro-Québec, who concluded that commercial availability of such units is now in sight. However, the need to develop cheap, efficient electrocatalysts for oxygen reduction remains.

The next presentations discussed chemical problems encountered in the nuclear power industry. S. R. Hatcher (Atomic Energy of Canada Ltd., Pinawa) gave a general review covering the chemistry of established and novel nuclear fuels, heavy-water production, and reactor operation.

M. Tomlinson (Atomic Energy of Canada Ltd., Pinawa) described how advances in chemical knowledge can help to assure long-term containment of nuclear wastes in underground formations.

In conclusion, the conference indicated the diverse nature of the chemical research and development which is required in order to benefit from Canada's abundant energy sources.

Chemical Institute of Canada,
Manitoba Section

September 25, 1978

M. TOMLINSON,
T. E. RUMMERY,
D. F. TORGERSON,
A. G. WIKJORD

M. Tomlinson, *Editor*

T. E. Rummery,
D. F. Torgerson,
A. G. Wikjord, *Associate Editors*

Sponsoring Organizations

for the

Chemistry for Energy Symposium

Sponsors: Alberta Energy Co. Ltd.
The Chemical Institute of Canada
Canada Department of Energy, Mines
    and Resources
Gulf Oil Canada Ltd.
Imperial Oil Ltd.
Shell Canada Ltd.

Keynote Address

Fluid Fuels—The Chemists' Problem

PETER J. DYNE

Department of Energy, Mines and Resources, Ottawa, Ontario

Firstly, consider some generalities of the energy problem. These break into three separate parts: the need to restrict or control the increase in energy consumption, the need to provide energy in new forms as a substitute for fluid fuels and the extent to which other energy sources can provide this replacement.

These three aspects are summarised in Figures 1, 2, & 3. Without getting into an argument of how much energy consumption can be controlled or reduced, the diagrams indicate the need for large new sources of energy over the next fifty years, even if energy consumption were to be significantly reduced.

These diagrams imply by the choice of a single energy axis that all forms of energy are equivalent where, in practice, they are not. When we supply energy in different forms in our national energy system we have to have new or modified devices to use that energy.

For example: - solar heat means houses with new designs and new structures;
- methanol as an automobile fuel means different engines and a new fuel distribution system;
- electricity for the transportation sector means electric cars and presumably development of batteries.

All our equipment requiring energy will have to be optimized for a new economic environment where energy or fuel, is not cheap. Sharply increased fuel costs will provide the incentive for more efficient energy using devices which, themselves may be more expensive in terms of capital than today's equipment. In the future then we have to pay as much attention to the energy using technologies as on the energy supply technologies.

There appear to be a number of ways in which this energy system can be rebuilt. Energy is not an end in itself. We are supplying energy for another more general purpose, - to provide society with the necessities and the luxuries of life. The way in which we use energy is determined in part by the sort of social and economic system we want and of course, the other way about;

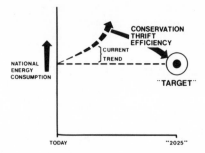

Figure 1.   Schematic showing the effect of energy conservation in reaching an arbitrary "target" for consumption

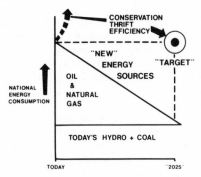

Figure 2.   Schematic showing the need to develop "new" energy sources to meet the "target" energy consumption

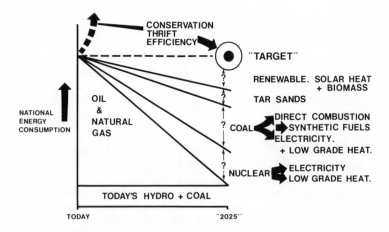

Figure 3.   Schematic illustrating the theme of the talk. How big can the various supply sectors be and how do they mesh together to provide energy in the required form?

how energy systems molded our social and economic systems.

We would like to have cheap energy, or, at least, energy as inexpensive as possible; we would like secure energy supplies (in the political sense) and we want to have supply systems with minimum environmental effects. Unfortunately, it is often difficult to get all three at the same time. For many countries, the cheapest supplies are for some time likely to be imported oil, these are not however, necessarily secure. The use of coal may be politically secure and relatively inexpensive but not environmentally sanitary. Synthetic fuels based on coal may pose environmental problems because of the large amounts of coal to be mined and processed, and may not be cheap either. Electricity, because it can be generated in a variety of ways and in many places responds well to security of supply. Environmental effects at the point of end use are minimal.

Electricity and the extent to which we build electrical energy economics is one of the central questions in planning our energy future. If you look at the ways in which we use energy, you find that almost all of them can or could be done via electricity. In a conceptual sense, one can imagine hydrogen fueled aeroplanes, the hydrogen being derived from the electrolysis of water. It is interesting to note that almost all of the future and futuristic energy sources, fission or fusion energy, photovoltaics, ocean thermal gradients, wave power, end up by producing electricity. One can argue then that we will have to build electrified energy economies.

One difficulty with electricity is that, it cannot be stored cheaply. It has to be made when it's used. Electrochemical storage, while very useful indeed, is expensive. A lead acid battery costing $45 stores less than 5 cents worth of electricity! Storage as chemical energy is, by contrast, extraordinarily convenient. Here 1 kWh thermal, as petroleum at $1/gallon, costs 1.2 cents/kWh(th); energy storage being thrown in free.

This little illustration brings us to the key role that chemists and chemical engineers have. However much electricity we may use there will be a great need for fluid fuels. Chemical energy storage is the best thing we have. With the fossil fuels we get the energy and the storage together, at bargain prices. However, once we start talking about synthetic chemical fuels we have to:-

1) Synthesise materials on a very large scale;
2) Supply the energy for the synthesis on the same large scale;
3) Do this in a sanitary fashion; and
4) Do all this "cheaply".

The fourth item is the catch: in a general sense we probably know how to do the first three. The central problem is one of cost or price.

We have built an energy machine which was designed to produce as if energy was free (energy even at today's prices is still extremely cheap). While we have become accustomed to significant increases in energy costs we still have to face the situation that they will be significantly higher in the future. While we may contemplate an increase in energy cost by a factor of say, two or three, without having a major effect on the way we do things, an increase of a factor of ten would make a very significant difference.

Let me give you some numbers which illustrate how "cheap" we have to be. As a standard, let us take today's "expensive" energy; a tar sands plant costing say two billion dollars. This produces 100,000 barrels/day. This two billion capital works out to $12.00 for 1kWh/day. That doesn't sound very much! Indeed, it isn't! To put this in another context, I am going to use this example to show how this idea of cost provides a stimulus for practical ideas. There is today rightly a very great deal of interest in storing and converting solar energy.

How much can we afford for solar devices? The average solar flux is about $1kWh/day/m^2$ so that for a solar conversion device at 100% efficiency we would like to have something costing, say $12/m^2$ or in the order of $1.00/square foot to be competitive with tar sands, etc. In case you don't know it, a thermopane window is likely to cost five or ten times that already! This, of course, is an unfair and oversimplified comparison: it ignores all sorts of things like operating and transmission costs. But even if we allow the solar conversion device to be ten times as expensive (say $100/m^2$) that would still be a cheap piece of hardware!

This example brings out immediately the potential of natural photosynthetic processes. Plants do this photochemical conversion for a living! For them, all the research and development on complex photocatalytic processes has been done and they have been highly optimized; the energy collectors, the leaves, a self-eventing membrane, are almost free! As a consequence, for the conversion of solar energy to a chemically stored energy, natural photosynthetic processes will be very hard to beat. They don't, of course, give us a fluid directly: we have to do something with the cellulose to make it into a more usable form, but that may be a lot easier than trying to reproduce the photosynthetic step.

Having said that I wouldn't want to discourage fundamental work in photocatalytic reactions, it is much too early to assess the true technical implications of such work and we have little idea of where it can lead. We can only say that it is working in the right sort of area.

Besides the question of cost there is the question of scale. In mass terms our present oil consumption, 2 million barrels/day, amounts to about 300,000 tons/day. (Little wonder that, on a ton/mile basis, oil and gas transport is the largest single commodity being moved around). When you start talking of synthetic fuels (or of semisynthetic fuels like oil from tar sands

or methanol from wood) one can see major logistical problems
in just handling and processing that amount of 'stuff' everyday.
One could conclude, that synthetic fuels will have to be restricted in use to the transportation sector where they are
uniquely valuable and that other energy sources have to do
everything else.

The amount of 'stuff' to be handled leads to the two main
problems in any large scale production of synthetic fuels, first
the size and hence the cost of the equipment and second, the
environmental problems associated with the mass flows.

The scale of the operation itself leads to the need for large
capital investments. Chemical reactors and plants are big and
hence expensive because amongst other things the reactions don't
go fast enough. If the reactions could be speeded up then the
size of the plant could be reduced. Catalysts and catalysis are
therefore a key area of fundamental chemistry for the energy
field.

These large mass flows are, themselves, the source of environmental problems. To describe the plant simply in terms of its
production capacity of 300,000 tons/day is an over simplification. In any such process there will be other process flows
which, in some instances may amount to much more than the simple
product flow. The amounts of trace materials, even at a few
p.p.m. in concentration, which can accumulate or which may have
to be released amount, in times of months or years, to large
quantities. We are concerned about the potential environmental
effects of these trace materials. In spite of this concern,
however, we have little hard knowledge of these effects, in part
because we know so little about the chemistry of small amounts
of trace elements and compounds in large amounts of water or air.
If we then are going to come to rational terms with the environment, knowing what can be safely released and what cannot, we
have to know a lot more about chemistry or trace concentrations.

In brief then my keystone message for chemists and chemical
engineers in the energy business: they have the key problem
which nobody else will tackle, making synthetic fuels at a price
we can pay and with proper understanding of environmental
problems.

RECEIVED September 25, 1978.

Fossil Fuels

Prospects for Coal Conversion in Canada

N. BERKOWITZ

Research Council of Alberta, 11315 87th Avenue, Edmonton, Alberta, Canada T6G 2C2

Since the late 1960s, and more especially since 1973, when OPEC policies reversed fossil fuel pricing patterns that had virtually eliminated it as a major component of the Canadian energy economy, coal has not only regained substantial footholds in industrial fuel markets, but also attracted increasingly serious attention as a key resource from which, in future, more diverse energy demands could be met.

The beginnings of this renaissance can be traced to the early 1960s, when electric utilities in Alberta and Saskatchewan found it economically more advantageous to burn surface-mined coal in mine-mouth generating facilities than to fuel power stations in load-centres with natural gas; and by the early 1970s, these advantages, in conjunction with developing concerns about future natural gas prices, proved so persuasive that Alberta adopted coal-firing of new base-load thermal plants as provincial policy. But wider appreciation of benefits from greater reliance on coal came only with sharply escalating oil and gas prices, and with the recognition that known reserves of oil and gas in the Western Canadian sedimentary basin will not sustain demands for oil and natural gas beyond the late 1980s. These factors are now tending to accelerate re-entry of coal into some of its former traditional Western Canadian markets, and beginning to remove obstacles to the use of coal where coal costs are very much higher than in the prairie provinces: Nova Scotia, where steps are now being taken to reduce dependence on offshore fuel oils by greater utilization of indigeous underground-mined coal, is a case in point.

Even if current (per million btu) price differentials between coal and gas (or oil) do not widen much further, such direct substitution of coal for other hydrocarbon fuels is certain to become increasingly attractive - and may, indeed, prove imperative in the national interest. But in the long run equally important is that technological advances, coupled with the abundance and projected cost of Western Canadian coal, now make it possible to contemplate large-scale conversion of coal into gaseous and/or

liquid hydrocarbons, and thereby augment production from remote
arctic regions and manufacture of "synthetics" which will be
needed to offset diminishing supplies of conventional oil and
gas from more "traditional" sources.

The Chemistry of Coal Conversion

In chemical terms, the simplest conversion technique is the
transformation of coal into a combustible gas by *gasification*.
In its earliest form - introduced in Britain ca. 1860 by Sir
William Siemens - this involved generation of a producer gas,
mainly a mixture of CO, CO_2 and nitrogen, by incomplete combustion of coal in air, i.e., by

$$C + O_2 \longrightarrow CO_2 \qquad \cdots \qquad (i)$$

followed by

$$CO_2 + C \longrightarrow 2CO \qquad \cdots \qquad (ii)$$

The yield and/or heat value of this gas was later improved by
co-generation of a socalled water gas via

$$C + H_2O \longrightarrow CO + H_2 \qquad \cdots \qquad (iii)$$

by alternating (cyclical) injection of air and steam into the
burning fuel bed.

In modern gasification practice, principal reliance is
placed on the carbon-steam reaction (iii) which, depending on
the mode of operation of the reactor, may be variously accompanied by the "shift" reaction

$$CO + H_2O \longrightarrow CO_2 + H_2 \qquad \cdots \qquad (iv)$$

as well as by carbon hydrogenation

$$C + 2H_2 \longrightarrow CH_4 \qquad \cdots \qquad (v)$$

and thermal cracking of pyrolytically generated volatile matter,
formally represented by

$$C_mH_n \longrightarrow \frac{n}{4}CH_4 + \frac{m-n}{4}C \qquad \cdots \qquad (vi);$$

and partial combustion serves primarily as a source of process
heat. In some experimental systems, combustion in the gasifier
is accordingly replaced by an externally generated liquid (1) or
solid (2) heat-carrier which is circulated through the gasifier.

Where such external heat-carriers are employed, and the coal
is gasified by injection of steam only, it is obviously immater-
ial whether air or oxygen is used for generation of the heat

source. But when process heat is supplied by partial combustion
in the gasifier, the choice between air and oxygen is critically
important: If combustion is sustained with air, the product gas
will contain a large proportion of nitrogen and consequently only
have a heat value of $\sim$120-150/btu/scf ($\sim$4.5-5.5 MJ/m^3), while
combustion with oxygen will yield a gas that typically contains
$\sim$80 v/v per cent CO + H$_2$ and, depending on the actual CO:H$_2$
ratio, possesses a heat value of 270-350 btu/scf ($\sim$10-13 MJ/m^3).

  This difference bears directly and importantly on how the
product gas can be used. Because of the impracticability of
separating nitrogen from it, the *low-btu gas* can only by deployed
as an industrial fuel; and since it requires a larger combustion
space than a richer gas, the facilities in which it could be used
must be specifically designed for it. In contrast, a *medium-btu
gas* from oxygen-blown reactors can be accommodated in existing
installations with only minor burner-tip adjustments; and after
clean-up (to remove CO$_2$, H$_2$S, COS, etc.) and correction of the
CO:H$_2$ ratio to the stoichiometric values needed for downstream
processing, it also offers a petrochemical feedstock (or "syngas")
fully equivalent to those now most often made by partial oxidation (or "reforming") of natural gas or naphtha.

  Adjustment of the CO:H$_2$ ratio is effected by the shift reaction (iv) which proceeds over a chromium-promoted iron catalyst
at 700-800°F (370-425°C) or over a reduced copper/zinc catalyst
at 375-450°F (190-230°C); and the fraction of crude gas sent
through the shift reactor is calculated from the initial gas composition and specific downstream requirements. The latter are
illustrated by

(a) methanation, i.e. CO + 3H$_2$ $\longrightarrow$ CH$_4$ + H$_2$O, used to
    produce a high-btu "substitute natural gas" (SNG)
    with 940+ btu/scf ($\sim$35+ MJ/m^3);

(b) methanol synthesis, i.e. CO + 2H$_2$ $\longrightarrow$ CH$_3$OH; and

(c) ammonia synthesis, i.e. N$_2$ + 3H$_2$ $\longrightarrow$ 2NH$_3$, in which
    case all CO is abstracted from the syngas.

  For production of liquid hydrocarbons and oxygenated compounds other than methanol, shifting is usually carried to CO:H$_2$
ratios in the range 1.8-2.4 and use is made of variants of
Fischer-Tropsch synthesis (3).

  Gas cleaning before and/or after shifting is accomplished by
absorbing acid gases in, e.g., hot aq. carbonate, aq. methyl-
amino-propionic acid, dimethyl-amino-acetic acid, mono- or di-
ethanolamine, dimethyl-ethers of polyethylene glycol or methanol
(at between -18° and -62°C). Proprietary techniques employing
these (or other) adsorbents are being routinely used in natural
gas processing and can reduce residual concentrations of CO$_2$ and
H$_2$S to well below 20 ppm and 5 ppm respectively.

The chemistry of *coal liquefaction* is less well understood;
and how studies of this matter are interpreted depends to some
extent on how the "molecular" structure of coal is perceived.
    What is evident is that introduction of sufficient hydrogen
into coal to raise its atomic H/C ratio from ~0.65 to >1.0 (in
"primary" coal liquids that can be upgraded by common refinery
procedures) is only possible in a relatively narrow temperature
range centered on 800°F (425°C). Only in this range does active
thermal decomposition generate "free radicals" that can be stabilized by hydrogen addition before they randomly repolymerize or
crack to extinction. At substantially lower temperatures, addition of hydrogen - e.g., by reacting coal with lithium in ethyl-
enediamine at 90-100°C (4) - succeeds only in increasing the solubility of the coal in amine-type solvents, even though as many
as 55 H atoms per 100 C atoms can be added in this manner; and at
temperatures much above 850°F (450°C), rapid concurrent carbonization (and consequent aromatization) of the coal makes hydrogenation progressively more difficult.
    Confirmation that conversion of coal into liquids depends on
limited pyrolytic disruption of coal "molecules" and on prompt
stabilization of the resultant fragments by hydrogen is provided
by liquefaction in a *hydrogen-donor* which allows such reactions
as

Kinetic studies (5) of such systems indicate that the initial
stages of liquefaction involve conversion of the coal into a more
or less completely pyridine-soluble solid and thereafter into a
benzene-soluble material which is gradually transformed into a
viscous liquid as increasing amounts of hydrogen combine with it.
This process can be catalyzed by, e.g., cobalt molybdate, but
proceeds rapidly even in the absence of catalysts. At 775°F
(400°C), total py-solubility (and ~60 per cent solubility in ben-
zene) can be attained within less than 10 minutes.
    A notable feature of liquefaction in H-donor systems is that
the effective life of the donor can be substantially prolonged by
conducting the reaction in the presence of molecular hydrogen.
But it is not yet clear whether this effect stems from direct hydrogenation of the coal by H_2 (and from consequent lower demand
on the donor) or from re-hydrogenation of the donor as it is
stripped of available hydrogen; and neither is much known about

how the coal decomposes or about what types of fragments are
most amenable to stabilization by hydrogen.

Uncertainty about the chemistry of liquefaction has, however,
not inhibited development of several "second generation" liquefaction processes that depend on H-transfer from a donor; and a
number of such processes have in fact reached advanced stages of
development. By simplifying operations, these offer important
technical and economic advantages over classic *Bergius hydrogenation* which has been used in Germany and Britain to manufacture
synthetic gasolines, diesel fuel and heating oils from coal and
coal tars in the 1930s and throughout World War II.

The Bergius approach entailed two-step processing, with a
coal-oil slurry first being reacted with H_2 over iron oxide or
NH_4Cl-promoted tin catalysts at 457-485°C/25-70 MPa, and the resulting "middle oil" (b.p. 180-325°C) then being upgraded by
vapour-phase hydrogenation over a tungsten sulphide catalyst (6).

The liquefaction techniques now being developed resemble this
form of hydrogenation in retaining a two-step sequence, but are
much more energy-efficient and also return better yields through
being less drastic. The first stage typically entails reaction
of coal with H_2 and a donor - usually a hydrogenated recycle oil -
at 370-450°C/10-18 MPa; and, in some versions, this stage provides
options for producing *solvent-refined coal*, i.e., a substantially
mineral matter- and sulphur-free solid fuel which also offers raw
material for manufacture of carbon electrodes and other specialty
products. In that case, hydrogen transfer to the coal is limited
to levels that allow the coal to dissolve (or disperse) in the
donor fluid, but do not induce concurrent liquefaction. The dis-
persion is then filtered, and the solute is separated from solvent by pressure-reduction, distillation, precipitation or a
combination of these.

The chemistry of a third group of conversion techniques -
i.e., *partial conversion* methods which skim hydrocarbon gases
and/or liquids from the coal and leave a char suitable for use as
a boiler fuel or gasification feedstock - is, if anything, even
more speculative than the chemistry of liquefaction.

Except for supercritical gas extraction (see below), these
techniques involve very rapid heating of the coal to temperatures
at which it decomposes, and utilize the fact that the coal will
then generate an amount of "volatile matter" that far exceeds the
nominal volatile matter content determined by standard analytical
procedures (7). Under optimum operating conditions, yields of
liquid hydrocarbons can therefore be pushed much beyond those
accruing from carbonization in coke ovens or (coal)-gas retorts.
For example, while conventional carbonization (at heating rates
<5°C/min) will, at best, furnish 25-30 gals (∿95-115 litres/
tonne) of tars and "light oils" per ton of (dry, ash-free) coal
material, flash pyrolysis (which heats the coal at several hundred degrees C/min to a maximum temperature near 575°C) can
yield 40-45 gals ton (∿165-175 litres/tonne).

To account for this behaviour, it is generally assumed that
the coal "molecule" breaks down much more drastically when very
rapidly raised to decomposition temperatures, and that repoly-
merization of the radical fragments is inhibited by their fast
discharge from the coal. Some support for this thesis can be
seen in the fact that, in a coke oven, "primary" tars will aroma-
tize the more the longer they are in intimate contact with the
hot coke. But a necessary corollary is, obviously, that the
fragments must be supposed to stabilize themselves in the vapour-
phase by internal disproportionation.

Supercritical gas extraction depends on the fact that the
vapour pressure of a solid or liquid can be greatly increased by
contacting it with a compressed gas, and that this enhancement is
the more pronounced the greater the gas density (8). In principle, it is consequently possible to transfer into the vapour
phase substances that are otherwise substantially non-volatile.

The technique has been successfully used for de-asphalting
petroleum fractions, and has now come under study as a means for
extracting thermally generated non-distillable coal liquids at
temperatures at which they suffer little further degradation (9).
With toluene or other light solvents at ∿400°C/10 MPa, it has
proved possible to extract up to 35 w/w per cent of coal material, and to recover the extract by lowering the pressure and
thereby reducing the vapour density.

Status of Coal Conversion Techniques

Sustained demand for fuel gas and petrochemical feedstocks
in countries where natural gas or naphtha was not readily available has occasioned progressive refinement of several gasifier
systems that were first introduced into commercial practice in
the mid-1930s and now have a proven record of performance in
numerous contemporary plants.

The most prominent of these are the *Lurgi* (10), *Koppers-
Totzek* (11) and *Winkler* (12) reactors which are currently used
for production of hydrogen in several ammonia plants, and the
Wellman gasifier (13) which provides on-site low-btu fuel gas
for various industrial installations in South Africa. Lurgi
generators are also used to manufacture syngas in South Africa's
well-known Sasol plant (14), which since 1955 has produced synthetic gasolines, diesel fuels and chemicals by Fischer-Tropsch
syntheses; and more recently, such units have come into operation as the front-end of a 190-MW prototype low-btu gas-fuelled
combined cycle generating plant in the German Federal Republic
(15).

In passing it might be observed that these gasifiers illustrate the wide design freedoms which coal gasification allows;
Wellman and Lurgi reactors operate with fixed beds and dry ash
discharge at, respectively, atmospheric pressure and 3.0-3.5
MPa (∿400-500 psia). The Winkler generator gasifies a fluidized

bed at atmospheric (or higher) pressures. And the Koppers-
Totzek gasifier, which operates at atmospheric pressure, uses
oxygen-entrained coal, with ash being released as slag.

But despite satisfactory performance of commercially proven
gasifiers, and the operational flexibility afforded by the variety of available reactor designs, there is concern that the
capacities of these units and/or the restrictions which they
place on their respective coal charges, may make them unsuitable
or too costly for very large gasification plants expected to be
needed in the future. The largest gasifiers presently available
- the Lurgi Mk. IV and the 4-Burner Koppers-Totzek - have throughput capacities of ∿800-850 tons/day, and a 250 MM scf/day (∿7 x
10^6 m^3/day) SNG plant would therefore require continuous operation of at least thirty such units. Massive R & D programs have
therefore been initiated, particularly in the United States, in
efforts to perfect a number of alternative "second generation"
systems. Mostly designed with an eye to eventual production of
high-btu (940+ btu/scf) SNG, these are intended to reduce costs
by offering mechanically simpler gasifiers which could be scaled
up to large capacities, operate at pressures of ∿6.9 MPa (∿1000
psia) in order to eliminate downstream gas compression, and/or
allow hydrogasification (which obviates subsequent gas shifting
and methanation). Examples are Bituminous Coal Research Inc.'s
Bi-Gas Process (16), the *Synthane Process* (17) which was originally conceived at the US Bureau of Mines, IGT's *HYGAS Process*
(18), and Kellogg's *Molten Salt Process* (1). Reports from
pilot-plants suggest, however, that some of the "second generation" gasifiers pose problems associated with balanced operation
of their component fluidized beds; and having regard to projected
needs in the 1980's, much of the practical interest has therefore
shifted to modified forms of proven gasifiers that promise to
lower costs.

One such modification is the *slagging* Lurgi gasifier, which
differs from the conventional "dry bottom" unit by operating at
higher temperatures and discharging ash in molten form. As a
result of design changes required for this mode of operation, it
has substantially greater throughput capacity per unit reactor
volume, and can reportedly accept coal types and sizes which the
dry-bottom Lurgi cannot handle. Present indications are that the
slagging Lurgi gasifier will be ready for commercial use in the
early 1980s.

A second is a *pressurized* version of the Koppers-Totzek
generator which, while retaining the advantages of a high-temperature, entrained-coal gasifier, would also promote formation
of methane (by reaction v). Developed by a Krupp/Shell consor-
tium, a 150 t/d commercial prototype is scheduled to undergo
extended tests in the German Federal Republic later this year.

And finally, mention must be made of the *Texaco* gasifier
(19) which has been adapted from an industrially proven oil gasification system and accepts a coal-water slurry. Although the

coal-fed version has so far only been operated in a 15 t/d test
facility at Montebello, Cal., Texaco is reportedly ready to guar-
antee its performance for commercial use; and recent estimates
suggest that, when used to produce hydrogen for ammonia synthesis,
it would be ~15 per cent cheaper than its more conventional coal-
based competitors.

   Liquefaction technologies and partial conversion processes
are generally less advanced than gasification methods. Bergius
hydrogenation is now only of historical interest, and none of the
second generation procedures has to date been taken beyond relatively small pilot plant testing. However, current emphasis on
developing some of these schemes makes it likely that at least
two or three alternative processes will be available for commercial deployment by the mid-1980s.

   Leading contenders appear to be Hydrocarbon Research Inc.'s
H-Coal Process (20), Exxon's *Donor Solvent (EDS)* Process (21),
and Gulf Oil's *SRC-II* (22). The latter is an outgrowth of Gulf's
Solvent-Refined Coal (SRC) Process (23) which is commanding much
attention in its own right as a means for converting high ash/
high sulphur coals into environmentally clean solid fuels (with
<0.5 per cent S, <0.1 per cent ash). With the exception of the
H-Coal Process and the Lummus Corp.'s somewhat similar *Clean
Fuels from Coal (CFFC)* Process (24), all these techniques, as
well as several variants - e.g., Gulf's *Catalytic Coal Liquids
(CCL)* Process (25) and Consolidation Coal Company's *Synthetic
Fuels (CSF)* Process (26) - entail reacting a coal-donor slurry
under molecular hydrogen at temperatures and pressures in the
range 400-450°C and 1-3 MPa respectively, and then hydro-treating
the "primary" liquids and spent donor. Differences between different processes relate mainly to specific operating conditions,
to catalysts (where such are used), and to the manner in which
the primary liquids are freed of unreacted coal and mineral matter before being taken to upgrading.

   The H-Coal and CFFC Processes are unique in the sense that
first-stage liquefaction is achieved by circulating the coal-
donor slurry through an "ebullated" (i.e., partly fluidized) bed
of uniformly sized catalyst pellets.

   Of the *partial conversion* techniques, the most immediately
interesting are FMC Corp.'s COED (Coal-Oil-Energy Development)
Process (27) - which, when combined with gasification of residual
char, is now known as the *COGAS* Process (28), and Occidental Oil's
Garrett Flash Pyrolysis Process (29). In the former, rapid heating to progressively higher temperatures is accomplished by suc-
cessively cascading the coal through three reactors against a
countercurrent stream of hot combustion gas. This yields up to
40 per cent of recoverable hydrocarbon gases and heavy oils (or,
more correctly, tars) which can be upgraded in much the same manner as the primary liquids from liquefaction. In Garret Flash
Pyrolysis, the coal is raised to the desired final temperature
in a single step by contacting it with hot char (from a combustor

where some coal or char is burned in air); and depending on
whether the process is operated near ∿870°C (∿1600°F) or ∿570°C
(∿1050°F), hydrocarbon gases or primary liquids - in either case
amounting to 30-35 per cent of the dry, ash-free coal substance
charged to it - are obtained.

Both the COED process and Garrett Flash Pyrolysis have been
"proved" in small demonstration plants; and subject to tests designed to establish optimum operating conditions for a particular
feed coal, both are claimed to be ready for commercial use.

The Influence of Coal Properties on Behaviour in Conversion Process

Because of the dissimilarities between different kinds of
coal, some care must be exercised to ensure compatibility between a feedstock and the equipment in which it is to be processed; and as in other technologies, optimum performance is
contingent on operating under appropriate conditions.

In *gasification*, the principal constraints arise from the
inability of some reactor configurations to accept caking coals
(which, when heated to temperatures above ∿350-400°C, form more
or less distended cokes that may cause blockages in the coal
delivery system) or, in the case of fixed bed gasifiers, coal
below a certain size. In some instances it is therefore necessary to "condition" the feed coal by pre-heating in air at ∿350-
400°C (to destroy caking propensities), and/or screening it to
remove undersize material before charging it to the gasifier.

In dry bottom units, a further impediment may arise from an
excessively low ash fusion temperature. This can be countered by
reducing the operating temperature, but since that also lowers
the gasification velocity, a proportionately smaller coal
throughput must then be accepted.

Beyond these matters, the most important coal properties are
(a) reactivity, which is, broadly speaking, a function of the
porosity and non-aromatic carbon content of the coal, and consequently varies in an inverse sense with the "rank" (or total
organic carbon content) of the coal, and (b) mineral matter and
sulphur contents. The rank-dependence of reactivity usually requires raising gasification temperatures from ∿1550°F (∿850°C)
in the case of lignites to ∿1850°F (∿1000°C) for more mature
high volatile and medium volatile bituminous coals in order to
make gasification proceed at commercially acceptable rates. And
large mineral matter and/or sulphur contents may create difficulties in the ash discharge system as well as require special
attention to adequate product gas cleaning.

In *liquefaction*, the additional properties of importance
are (a) the petrographic composition of the coal, and (b) its
oxygen content.

Of the three major groups of petrographic constituents
(termed macerals; see appended note), *vitrinites* and *exinites*,

which represent the bulk of the (organic) coal substance, can as
a rule be quite easily liquefied - though ease of liquefaction is
related to chemical and physical reactivity and tends to decrease
with increasing rank. (Preferred liquefaction feedstocks are,
consequently, coals up to and including high volatile bituminous
rank). However, *inertinites* will generally leave a liquefaction
reactor essentially unchanged with other solid matter; and what
fraction of a feed coal appears as primary liquids depends therefore on the proportion of inertinites in it.

Oxygen in coal represents an undesirably hydrogen sink
(since most will eventually appear in H_2O); but since oxygen
contents vary inversely with carbon contents, any selection of
coal for liquefaction involves some trade-off on this matter:
while high oxygen contents are, *a priori*, deleterious because
they increase overall hydrogen consumption per barrel of primary
coal liquids (and lower the yield per ton) using a coal with
relatively little oxygen is also tantamount to using one with
lower reactivity.

With respect to *partial conversion* by flash pyrolysis, the
principal consideration in a choice between otherwise equivalent
coals is the fact that liquid yields tend to increase with rank
up to high volatile bituminous coals and thereafter to fall off
sharply.

Having noted these technical matters, it should, however,
also be observed that the operational flexibility of available
processing equipment usually quite subordinates them to coal
costs (which represent between 25 and 40 per cent of the plant
gate costs of the product). Most future conversion plants are
therefore likely to use lignites and subbituminous coals which,
in Canada as in many other countries, are abundantly available
at low cost through surface-mining.

Canadian Opportunities for Coal Conversion

The technical status of conversion processes and relative
fuel costs make the extensive reserves of easily accessible coal
in the Western Plains key resources that could play an important
role in meeting future requirements of petrochemical feedstocks,
fuel gas and liquid fuels. And developed concurrently with oil
sands and "heavy" (Lloydminster-type) oil reservoirs, they could
allow Canada to regain virtual energy self-sufficiency within a
comparatively brief time-frame.

Because of the massive "unconventional" reserves of liquid
hydrocarbons afforded by oil sand bitumens and heavy oils,
Canadian interests in coal conversion are generally more likely
to centre on gasification than on liquefaction, and to focus on
long-term supply of fuel gas (which could in many cases be substituted for oil where coal can not, and thereby reduce projected
oil supply shortfalls).

Indicative of future directions is the attention now being given in Alberta to the feasibility of substituting a *medium-btu* coal gas for natural gas presently used as fuel in the two developing petrochemical centres (at Ft. Saskatchewan and Medicine Hat). By the early 1980s, by which time natural gas will, on present scheduling, have attained approximate price parity with oil (on an equivalent btu basis) at Toronto city gate, the coal gas is expected to offer substantial cost advantages.

If such a substitution program is implemented, a logical second step would be to use the same gas (after shifting and cleaning) as feedstock in some of these plants. The principal candidates would be methanol and ammonia plants which would require relatively little retro-fitting.

But much more important, in the long run, is that measures taken *locally* to displace natural gas on economic grounds could also serve to ensure economic supplies of fuel gas to residential and industrial consumers in other parts of Canada.

Notwithstanding the current surplus of natural gas in Western Canada (which is prompting demands for government consent to export additional volumes), there is broad agreement that deliveries from proved and expected future reserves in the Western Canadian sedimentary basin will not suffice to meet projected Canadian demands much beyond the late 1980s; and it is, in fact, for just this reason that gas exploration in arctic "frontier" regions is being stepped up. (In the Mackenzie Delta, some 6 Tcf (0.17×10^{12} m^3) of a potential ultimate 50-100 Tcf ($1.4-2.8 \times 10^{12}$ m^3) have already been delineated; and in the Polar Islands region, where ultimately recoverable volumes are also thought to range as high as 100 Tcf, approximately 12 Tcf (0.35×10^{12} m^3) have been proved.) However, for technical and environmental reasons, bringing such frontier gas to markets will prove costly, and consumer interests could be better served by offsetting future supply deficiencies via coal gasification. (It is, in this connection, worth noting that a pipeline system capable of gathering and transporting 1 Tcf (28×10^9 m^3) per year - the likely minimum rate for an economically viable operation - from the Arctic Islands to Southern Ontario would require a capital investment of some 8-10 billion dollars (in 1978 $$) and demand a gas price in the order of $4.50/Mcf ($15.90/100 m^3) at Toronto city gate.)

If it is assumed that any "synthetic pipeline gas" must *of necessity* match the heat value of natural gas, gasification may only be marginally attractive. Recent estimates (30) suggest that such substitute natural gas (SNG) would be cheaper than arctic gas - but, in the foreseeable future, still cost substantially more than natural gas from presently developed reservoirs; and to secure cost advantages over arctic gas, gasification would have to be conducted in very large plants, possibly using second generation technology that is not expected to be commercially available before the late 1980s. From a Canadian viewpoint,

having regard for requirements as well as for the geographic
distribution of resources and markets, it appears therefore preferable to abandon the notion that fuel gas must have a heat
value ∿1000 btu/scf (∿37 MJ/m^3); to recognize that even a gas
with as little as 700-750 btu/scf (∿26-28 MJ/m^3) could be readily burned in existing equipment; and to make use of medium-btu
gas *per se* by rolling it into natural gas supplies. Since the
major part of the cost of SNG is associated with syngas cleaning,
shifting and methanation, medium-btu gas would, in 1980, be pro-
ducible at significantly lower plant gate prices than natural gas
is then expected to command at the wellhead; and a 2:1 mixture of
natural and medium-btu gases with ∿750-780 btu/scf (∿28-29 MJ/m^3)
could be delivered to Central and Eastern Canadian markets at no
greater per MM btu cost than natural gas itself.

Moves in this direction would allow gasification to be very
gradually phased in as demand arises, and to employ proven reactor systems (which do not pose the risks associated with large
"second generation" facilities). And bearing in mind the very
large reserves of surface-mineable (let alone other) Western
Plains coal surplus to the long-term needs of Alberta and Saskatchewan, gas supplies from the Western sedimentary basin could
in this manner be virtually doubled. The consequent indefinite
deferment of development of arctic reservoirs would also serve
to hold escalation of fuel gas prices to more manageable rates
than must otherwise be expected.

Prospects for coal *liquefaction*, to which little attention
has so far been paid in Canada, are, at this point in time, less
clearly perceivable.

Much of the impetus for development of liquefaction technology in the US appears to derive from a need to convert Eastern
high-sulphur coals into environmentally acceptable boiler fuels;
and in Canada, where the overwhelming bulk of coal reserves (>90
per cent) is comprised of Western coal with, generally, less than
0.75 per cent S, this is not a matter of concern. However, liquefaction could in the longer run prove an attractive means for
augmenting "synthetic" oil supplies from other indigenous sources.
As matters stand, remaining proved reserves of conventional petroleum (almost all in Alberta) total 6.34 x 10^9 bbl (1.008 x
10^9 m^3) and supplies of "synthetic" light crudes from oil sands
bitumens are not expected to exceed 600-700 bbl/d (95.5-112 x
10^3 m^3/d) by 1990 or ∿1.25 x 10^6 bbl/d (0.2 x 10^6 m^3/d) by 2000
A.D. Even if supplemented by upgrading of other heavy oils,
total deliveries in the early 1990s are therefore projected to
leave a deficit of 0.5-1.0 x 10^6 bbl/d (80-160 x 10^3 m^3/d) that
would have to be closed by imports. (The possibility that this
gap can be significantly narrowed by production from "frontier"
reservoirs (such as the Beaufort Sea, where extensive "potentially favourable" sediments resembling productive oil-bearing strata
elsewhere are believed to exist) seems remote. Apart from the
fact that exploration has so far failed to find commercially

significant oil occurrences there, bringing a fully proved reservoir of this type on stream would require a lead period of
at least 10-12 years.)

Modest further escalation of international oil prices and/
or a lowering of currently indicated liquefaction costs through
further R & D could, in these circumstances, make coal liquefaction an important tool for minimizing, if not effectively
eliminating, otherwise extensive dependence on foreign sources
of supply. What ought to be borne in mind here is that a price
differential favouring offshore "conventional" petroleum over
indigenously produced "synthetic" oil could be greatly reduced
by appropriate fiscal policies respecting the internal cash flow
which new (direct and indirect) employment opportunities created
by "synthetic" oil production would generate.

However, with respect to oil from coal, two other possibilities - both less capital-intensive and less productive than a
full-scale liquefaction facility, but initially perhaps more
easily fitted into Canada's energy economy - present themselves.

The first, which would link existing and projected thermal
generation of electric energy to production of liquid hydrocarbons from coal, would involve subjecting the coal to, e.g.,
Garrett flash pyrolysis before sending the residual char to the
boilers. Under optimum operating conditions, this would yield
$\sim$1.8 bbl/t of dry, ash-free coal material (315 litres/tonne)
charged to the process, or $\sim$1 bbl/t (175 litres/tonne) if coal
burned to generate the heat carrier for pyrolysis is taken into
account; and combustion trials have shown that the devolatilized
char, which still retains at least 10 per cent volatile matter
(as determined by standard laboratory tests) will burn as well
as the raw coal.

Since only relatively simple equipment is required, it should
prove feasible to flash-pyrolyze the coal at the generating
station or at the mine at which it is produced, and to transport
the primary liquids from several operations to a large central
upgrading facility. In Alberta alone, where coal consumption by
electric utilities is expected to climb from an annual rate of
$\sim$6.5 million tons in 1977 to 16-17 million tons in 1985 and $\sim$40
million tons in 2000 A.D., some 15 million bbl (2.4×10^6 m^3) of
a marketable "synthetic" crude oil in 1985, and at least 35 million bbl (5.6×10^6 m^3) in 2000 A.D., could be recovered in this
manner. Similar quantities could be produced in Ontario, with
additional volumes coming from Saskatchewan and Nova Scotia; and
processing costs would in all cases be partly offset by the substantially higher net calorific value of the chars taken to the
boilers.

The second option worthy of consideration is a linkage of
certain petrochemical operations (specifically, ammonia and/or
methanol production) to manufacture of liquid hydrocarbons by
Fischer-Tropsch or Kölbel-Engelhardt syntheses.

As presently projected, production of ammonia and methanol from coal-based syngas, although viable even in 1200 t/d plants, is limited by relatively small domestic requirements, and unless demand for *fuel-grade* methanol (as a component of automotive fuels develops, substantial immediate additions to existing capacity would only be warranted if access to firm export markets in the mid-western and western United States could be gained. However, at least two new ammonia plants with a combined capacity of 3000-4000 t/d are expected to be needed before the late 1980s to meet increased Canadian consumption of nitrogenous fertilizers and replace facilities then nearing the end of their economic life, and these could conceivably be developed for co-production of methanol, gasoline, diesel fuels and/or fuel gas.

If such operations employed Fischer-Tropsch technology, they would in effect, be scaled-down versions of South Africa's Sasol plant, with principal benefits accruing from economies of scale (in the gasification section) as well as from some contribution to Canadian oil (or oil products) output and from an in-plant ability to accommodate transient demand fluctuations by switching surplus syngas to one or another of the downstream synthesis units. But the more challenging, and potentially more attractive, alternative would be the extraction of carbon monoxide from syngas streams destined for ammonia synthesis by, e.g., membrane permeation techniques (which would replace gas shifting and, by avoiding $CO \rightarrow CO_2$, make more efficient use of carbon input into the plant), and production of hydrocarbons from CO and steam over iron at $\sim$500-600°F/115-150 psi ($\sim$250-300°C/800-1000 kPa). First reported by Kölbel & Engelhardt (31) and formally represented by

$$3CO \ + \ H_2O \longrightarrow (-CH_2-) \ + \ 2CO_2$$

this reaction has been shown to achieve virtually 100 per cent theoretical conversion of CO to hydrocarbons, with three-quarters of the total yield being comprised of C_5- C_{10} hydrocarbons (32). The only by-product of this synthesis is (environmentally acceptable) carbon dioxide which, if so desired, could in part be used to generate additional volumes of CO by cycling it through incoming raw coal and making use of the Boudouard reaction.

Unlike Fischer-Tropsch technology, Kölbel-Engelhardt synthesis is, of course, still experimental and requires further pilot-plant development before its technical and economic feasibility can be assessed. The same is true of $CO-H_2$ separation which, ideally, would have to be conducted at temperatures that conserve the sensible heat of the incoming syngas without excessively slowing gas migration. However, these and other conversion topics touched on in this review are sufficiently relevant to Canada's future energy economy to merit much more attention - and very much more concerted *action* than has so far been devoted to them. It is regrettable that the monies provided by Canadian governments for "paper

studies" of coal conversion and related matters still greatly
exceed the funds available for practical research and engineering development - and that the latter are, in fact, still almost
vanishingly small when compared with expenditures on other, no
more promising, energy programs.

Note: These (maceral) constituents can be identified and quantitatively measured by examining thin sections or polished surfaces
under a microscope, and reflect the nature of the primordial
source material as well as the conditions under which it was deposited. *Vitrinites* derive from humic gels, wood, bark and cor-
tical tissues; *exinites* are the remains of fungal spores, leaf
cuticles, algae, resins and waxes; and *inertinites* comprise unspecified detrital matter, "carbonized" woody tissues and fungal
sclerotia and mycelia.

Summary

Rapid escalation of natural gas and oil prices has not only
once again made coal an economically attractive industrial fuel
per se, but also brought it into focus as a resource from which,
in future, petrochemical feedstocks, fuel gas and synthetic
liquid hydrocarbons could be produced.

This paper touches on the chemistry of coal gasification
and liquefaction; comments on the current status of conversion
processes and the influence of coal properties on coal performance in such processes; and examines the contributions which
coal conversion could make towards attainment of Canadian energy
self-sufficiency. Particular attention is directed to a possible
role for the medium-btu gas in long-term supply of fuel gas to
residential and industrial consumers; to linkages between partial
conversion and thermal generation of electric energy; and to co-
production of certain petrochemicals, fuel gas and liquid hydrocarbons by carbon monoxide hydrogenation.

Literature Cited

1. Cover, A. E., Schreiner, W. C., and Skapendas, G. T.;
    Chem. Engng. Prog., (1973), 69 (3);
    Proc. Clean Fuels from Coal Symp., Inst. Gas Technol.,
    Chicago, Ill., (1973) Sept.
2. Curran, G. P.; Chem. Engng. Prog., (1966), 62 (2);
    Fink, C. E., Curran, G. P., and Sudbury, J. D.; Proc. 7th
    Synth. Pipeline Gas. Symp., Chicago, Ill., (1975) Oct.
3. Storch, H. H., Golumbic, N., and Anderson, R. B.:
    "The Fischer-Tropsch and Related Syntheses", Wiley & Sons
    (New York), 1951.

4. Reggel, L., Raymond, R., Friedman, S., Friedel, R. A., and
Wender, I.; Fuel, (1958), 37, 126;
Reggel, L., Raymond, R., Steiner, W. A., Friedel, R. A., and
Wender, I.; Fuel, (1961), 40, 339.
5. Neavel, R. C.; Fuel, (1976), 55, 237.
6. Krönig, W.; "Katalytische Druckhydrierung", Springer
(Berlin), 1950.
Donath, E. E.; in "Advances in Catalysis", (eds. Franken-
burg, Komarewski & Rideal), Academic Press (New York),
1956, vol. 8.
7. cf. ASTM Standards, D 3175-77.
8. Paul, P.F.M., and Wise, W. S.; "The Principles of Gas
Extraction", M & B Monograph CE/5, Mills & Boon (London),
1971.
9. Whitehead, J. C., and Williams, D. F.; J. Inst. Fuel, (1975),
49, 182.
10. Rudolph, P.E.H.; Proc. 4th Synth. Pipeline Gas. Symp.,
Chicago, Ill., (1972) Oct.
Elgin, D. C., and Perks, H. R.; Proc. 6th. Synth. Pipeline
Gas Symp., Chicago, Ill., (1974) Oct;
Lurgi Express Info. Brochure No. 1018/10.75, (1975).
11. Farnsworth, J. F., Leonard, H. F., Mitsak, D. M., and
Wintrell, R.; Koppers Co. Publ., (1973), Aug.;
Wintrell, R.; Amer. Inst. Chem. Engrs., Nat. Meet., Salt
Lake City, Utah, (1974) Aug.
12. Banchik, I. N.; Davy Powergas Inc. Publ. (1974);
Symp. on Coal Gasification & Liquefaction, Pittsburgh, Pa.,
(1974).
13. McDowell-Wellman Engng. Co., Brochure No. 576;
Hamilton, G. M.; Cost. Engng., (1963) July, 4.
14. Linton, J. A., and Tisdall, G. C.; Coke & Gas, (1957), Oct.,
402; (1957) Nov., (1958) April, 148.
15. Puhr-Westerheide, H., and Marshall, H.W.S.; Fall Meet.,
Canad. Electr. Assoc., Montreal, (1974) Sept.
16. Glenn, R. A., and Grace, R. J.; Proc. 2nd Synth. Pipeline
Gas Symp., Pittsburgh, Pa., (1968);
Grace, R. J.; Clean Fuels from Coal Symp., Inst. Gas.
Technol., Chicago, Ill., (1973) Sept.
17. Forney, A. J., and McGee, J. P.; Proc. 4th Synth. Pipeline
Gas Symp., Chicago, Ill., (1972) Oct.;
Carson, S. E.; Proc. 7th Synth. Pipeline Gas. Symp.,
Chicago, Ill, (1975) Oct.
18. Schora, F. C., Lee, B. S., and Huebler, J.; Proc. 12th
World Gas Conf., Nice, France, (1972) June;
Lee, B. S.; Proc. 5th Synth. Pipeline Gas Symp., Chicago,
Ill., (1973) Oct.;
-; Proc. 7th Synth. Pipeline Gas Symp., Chicago, Ill.,
(1975) Oct.

19. Child, E. T.; Symp. on Coal Gasification & Liquefaction, Univ. Pittsburgh School of Engng., Pittsburgh, Pa., (1974) Aug.
20. US Dept. of Commerce; NITS PB-234 203, (1974), pp. 254 et seq.;
Coun, A. L., and Corns, J. B.; "Evaluation of Project H-Coal:, US Dept. of the Interior, Office of Coal Research, Washington; PB-177 068, (1967).
21. Furlong, L. E., Effron, E., Vernon, L. W., and Wilson, E.L.; Chem. Engng. Progr., (1976) Aug.
22. Schmid, B. K., and Jackson, D. M.; 3rd Ann. Conf. on Coal Gasification and Liquefaction, Univ. Pittsburgh School of Engng., Pittsburgh, Pa., (1976) Aug.
23. Pastor, G. R., Keetley, D. J., and Naylor, J. D.; Chem. Engng. Progr., (1976) Aug.;
Wolk, R., Stewart, N., and Alpert, S.; EPRI Journal, (1976) May.
24. Sze, M. C., and Snell, G. J.; Proc. Amer. Power Conf., (1975), 37, 315.
25. Chung, S.; Conf. on Materials Problems and Research Opportunities in Coal Conversion, Ohio State Univ., (1974), 263.
26. Gorin, E., Lewbowitz, H. E., Rice, C. E., and Struck, R. J.; Proc. 8th World Petrol. Congr., Moscow (USSR), (1971).
27. Shearer, H. A.; Chem. Engng. Progr., (1973), 69 (3), 43.
28. Eddinger, R. T., and Sacks, M. E.; 79th Nat. Meet., Amer. Inst. Chem. Engrs., Houston, Tex.,(1975) March;
Paige, W. A.; 5th Synth. Fuels from Coal Conf., Oklahoma State Univ., (1975) May.
29. Sass, A.; 6th Ann. Meet., Amer. Inst. Chem. Engrs., New York, (1972) Nov.;
McMath, H. G. Lumpkin, R. E., and Sass, A.; 66th. Ann. Meet., Amer. Inst. Chem. Engrs., Philadelphia, Pa., (1973) Nov.;
Adam, D. E., Sack, S., and Sass, A.; ibid., (1973).
30. See, for example, the Lummus Corporation's 1977 technical and economic evaluation of the Synthane Process. This study was commissioned by the US Dept. of Energy.
31. Kölbel, H., and Engelhard, F.; Brennst. Chem., (1951), 32, 150;
ibid., (1952), 33, 13.
Kölbel, H., and Vorwerk, E.; Brennst. Chem., (1957), 38, 2.
32. Maekawa, Y., Chakrabartty, S. K., and Berkowitz, N.; 5th Canad. Symp. Catalysis, Calgary, Alberta, (1977) Oct.

RECEIVED September 25, 1978.

3

The AOSTRA Role in Developing Energy from Alberta Oil Sands

R. D. HUMPHREYS

Engineering and Managing Consultant, AOSTRA, Edmonton, Alberta, Canada

The oil sands are located in the northern section of Alberta and up until a year ago it was believed that they contained just over nine hundred and fifty billion barrels of in-place oil. This compares to the total Canadian reserves of conventional oil of approximately eight billion barrels. Outtrim and Evans (1) of the Energy Conservation Board of Alberta presented a paper just about a year ago and suggested that the reserves in the oil sands and the carbonates are considerably higher than the previously accepted totals.

The Prize

It is estimated that the total oil sands and carbonates may contain in the order of two and a half trillion barrels of in-place oil. Not all of this is recoverable. Less than 5% of the Athabasca deposit is recoverable by the mining methods using today's technology and perhaps less than 30% of the remaining oil in all the oil sands deposits is recoverable by in-situ techniques. This means however, that a total of about 300 billion barrels is potentially recoverable. This compares very favourably with the 8 billion barrels estimated in our Canadian conventional oil reserves. In years of supply it equates to 400 to 500 years of reserves - a very attractive prize.

The presence of the oil sands has been known for many years and attempts to exploit the deposits commercially go back as far as the turn of the century. These attempts included such endeavours as those by Bitumount, Abasand and others. Dr. Karl A. Clarke, who worked on tar sands over a period of many years with the Alberta Research Council, successfully developed the hot water process used in the GCOS and Syncrude operations. Currently the Great Canadian Oil Sands Ltd. plant is operating at design capacity of 45,000 BPCD and Syncrude Canada Ltd. is in the process of starting-up its operation north of Fort McMurray.

AOSTRA

During the mid 1970's the Provincial Government of Alberta concluded that the huge deposits were not receiving the proper attention merited for such a massive prize. They decided to do something about it and established a crown corporation called the Alberta Oil Sands Technology and Research Authority, now commonly referred to as AOSTRA, with initial funding of $ 100,000,000. The funding has since been raised to $ 146,000,000, to enable continued funding of some longer range projects, which will be discussed later in the paper. The objective of the Authority was to develop economically and environmentally acceptable technology through collaborative action by industry, university and government. This major objective was subdivided into two categories, musts and wants, each of which included three goals.

Musts.   A   At least one in-situ recovery process for each major oil sand reservoir type.
            B   Resolution of major technical problems of current surface mining technology.
            C   More effective and efficient upgrading technology.
Wants.   A   Evolutionary increases in percent recovery from in-situ deposits.
            B   Alternative extraction technology.
            C   Conversion of oil sand into higher valued petroleum and mineral products.

Method

The method by which AOSTRA sought to carry out its objectives was through collaborative action by industry, university and government by:

            A   Arriving at a consensus of major technological barriers.
            B   Having access to and organizing the technological base data.
            C   Taking full advantage of university research capability.
            D   Full participation of the Alberta and Canadian consulting industry.
            E   Joint industry/AOSTRA project evaluation and participation.

In-situ Technology. The initial major objective of AOSTRA then was to develop a new in-situ technology or speed up existing in-situ technologies for each of the four major deposits. A call for proposals resulted in response of 21 major applications and from these an in-situ program was established which included four tests in three of the four major deposits. They include:

            BP in the Cold Lake deposit.
            Shell in the Peace River deposit.

AMOCO in the Athabasca deposit.
NUMAC in the Athabasca deposit.
The AMOCO project is furthest advanced and began operations
in the latter part of 1977. The project was plagued by cold
weather problems during part of the winter period but the operation is now underway and some very encouraging results are coming
forward.

The BP project in the Cold Lake deposit is now completed and
operations just underway. Construction is progressing on the
Shell Oil project and is expected to be completed in mid 1979.
Initial field tests have been completed on the NUMAC project and
results are being assessed in order to determine the further
course of action.

Of all the fluids investigated AOSTRA has concluded that
steam and air are the only two that are cheap enough to inject at
high volumes to aid in recovery. These two fluids, either alone
or in combination, have been selected for testing in each of the
deposits.

The sequence of steps in an in-situ project requires a
lengthy period of time. If laboratory research is sufficiently
encouraging a field test can then be attempted. If the concept
works on the initial field test, usually at a cost of 1 - 3
million dollars, a full pattern test, involving 5 - 9 wells can
be carried out. If the pattern test proves successful technically and economically, it is then theoretically possible to move
to a full commercial operation. However, for a number of reasons
involving the variability of the deposit and the risks involved,
the next phase is usually in the form of a large prototype including a number of adjoining patterns. Large capital expenditures are required in this phase and it is hoped that the technology has been developed to the point where AOSTRA need not be
involved in this stage.

Mining Improvements. The second major objective was to develop
improved surface mining technology. This program seeks to eliminate some of the major short-comings of the current mining
projects. The chief of these being the problem of sludge production from the hot water process and the attendant problems of
sludge disposal, dam building, solids materials handling and land
reclamation. To this end AOSTRA is engaged in the search for a
dry extraction process. A successful process would recover the
8 - 10% bitumen presently lost to the tailings pond and would
eliminate the need for tailings ponds completely. A pilot plant
is currently being operated in Calgary based on a retorting
process developed by a Calgary engineer. If the initial test
work is successful, it is planned to transport the unit to Fort
McMurray for further evaluation.

The Lurgi-Ruhrgas process is a direct retorting operation
which has been proposed for use on the Alberta oil sands. It is
the view of the Authority that the next step in the further
development of the Lurgi process is to carry out a test run on

Athabasca oil sands using an existing German pilot plant and
negotiations are currently underway with Lurgi for such a test.
If the Authority decides to take further steps on either or both
of these two processes, it will be approaching industry to obtain
both financial and technical assistance.

Upgrading Bitumen. The third major objective was to develop more
effective and efficient upgrading technology. Several bitumen
upgrading processes have been considered for oil sand use, but
coking processes are being employed for the first two commercial
plants. The flexicoking process, in its first year of commercial
operation in Japan, has the advantage of gasifying excess coke to
a low BTU gas. Both catalytic and thermal hydrocracking processes are also being evaluated with pilot work underway by the
Federal Government.

Underground Mining. The final technology area considered is
underground mining of the oil sands. A number of underground
mining proposals were received during the past two years, and
after an extensive review of these with the help of a mining
panel, it has been concluded that there is not sufficient basic
data on the underground characteristics of oil sands on which to
base a decision. A number of initiatives have been taken to
gather this missing information. One example is the measurement
of some underground parameters during the construction of a 14
foot diameter tunnel south of Fort McMurray. Shaft sinking is
being considered and an estimate of the cost of sinking a shaft
through the overburden to the oil sands has been completed. A
chair has been established at the University of Alberta which
will address itself to the study of geotechnical properties of
the oil sands and the Authority has participated with a number
of companies in evaluating the Russian underground thermal
mining operations. The Board is currently assessing all of the
data and information received during this past year and is
formulating a plan for future action.

AOSTRA has appointed a professor at each of the three
Alberta Universities and has also awarded a number of fellow-
ships and scholarships for oil sands research. This is in
addition to specific contract research being carried out at the
universities. The contract research work has been focused on
those topics which support the field projects referred to earlier.
Industry/University seminars have been held on an annual basis,
the third one was held in Calgary in March, 1978.

Technology Ownership

The technology that is developed through any of the AOSTRA
programs is owned entirely by AOSTRA and AOSTRA becomes the sole
licensing agent in Canada. The license fee will be established
by agreement with the participating company or companies and if
agreement cannot be reached on a fair market value by this means,
an arbitration procedure has been established.

Funding

Most of the company partnered projects are funded on a
50:50 basis with shared management and control and the licensing
income is divided on a pro rata basis. The company may however
invite up to three other industry participants provided the
initial company is responsible for any new company default. The
partner companies' prior technology can be incorporated into the
total licensing package.

Summary

AOSTRA now has four in-situ field projects underway, has
completed an initial underground field test in the tunnel at Fort
McMurray, is constructing one dry extraction pilot plant, has
completed one upgrading pilot test on flexicoking and is currently
reviewing programs for Lloydminster type heavy oil recovery and
oil upgrading. Finally, as back-up to the field and pilot
programs, AOSTRA has been placing considerable emphasis on explor-
atory research at universities and the Alberta Research Council.

Literature Cited

1 Outtrim, C.P. and Evans, R.G., Alberta's oil sands reserves
        and their evaluation: Heavy Oil Symposium, 28th Annual
        Technical Meeting of the Petroleum Society of the Canadian
        Institution of Mining and Metallurgy, May 1977, 41 p.

RECEIVED September 25, 1978.

Microemulsions as a Possible Tool for Tertiary Oil Recovery

JACQUES E. DESNOYERS, REJEAN BEAUDOIN, and GERALD PERRON

Department of Chemistry, Université de Sherbrooke, Sherbrooke, Quebec, Canada J1K 2R1

GENEVIEVE ROUX

Laboratoire de Thermodynamique et Cinétique Chimique, U.E.R. Sciences, B.P. 45, 63170 Aubière, France

The primary oil recovery takes advantage of the pressure exerted by the natural gases which forces the oil through the wells. When this pressure drops, it can be built-up again by water flooding. Unfortunately, after these primary and secondary processes, there still remains up to 70% of the oil adsorbed on the porous clays. Consequently, in recent years, there have been tremendous efforts made to develop tertiary oil recovery processes, namely carbon dioxide injection, steam flooding, surfactant flooding and the use of microemulsions. In this latter technique, illustrated in Fig. 1, the aim is to dissolve the oil into the microemulsion, then to displace this slug with a polymer solution, used for mobility control, and finally to recover the oil by water injection (1).

Nature of Microemulsions.

Microemulsions are rather complex mixtures of water, oil, surfactant, cosurfactant, usually alcohols, and often other additives such as electrolytes which, when added in the right proportions, form spontaneously a transparent or translucid liquid. One of the most important features of these microemulsions is the large quantity of oil that can be dissolved or dispersed in it. It is primarily for this reason that they are used to such a large extent commercially as water soluble waxes, cutting oils, wetting agents, herbicides and pesticides, synthetic blood, etc. (2). Despite the obvious importance of these systems, with the notable exception of Schulman *et al.* (3), few fundamental studies have been made until fairly recently. Still now strong disagreement exists amongst authors on the origin of the stability and on the structure of microemulsions. Following the original suggestions of Schulman, many consider microemulsions as a special case of emulsions where the small size of the droplets comes from the formation of a mixed film having a near zero or negative interfacial tension (3). Others, following the school of Friberg (4), prefer to consider the microemulsion as a swollen or inverse micelle. Their evidence comes mainly from very systematic studies of phase diagrams which indicate that the microemulsions are in fact extensions of the normal micel-

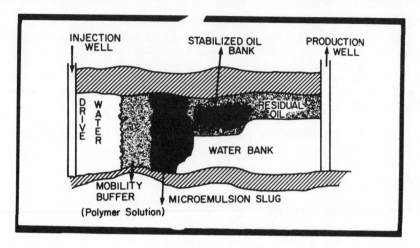

Figure 1.   Oil extraction by using microemulsions (1)

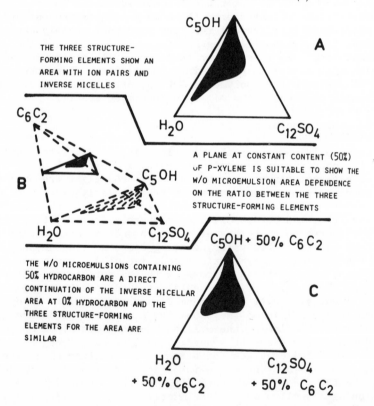

Figure 2.   Phase diagram of a water–pentanol–sodium dodecylsulfate-p-xylene microemulsion (4)

lar phase, as shown in Fig. 2.

As is often the case, we have become involved in microemulsions somwehat by accident. In the last five years or so we have
been making systematic studies of the thermodynamic properties of
aqueous organic mixtures and of electrolytes in these mixed solvents. Of particular interest were our heat capacity measurements.
With a differential flow microcalorimeter $(\underline{5},\underline{6})$ it is possible to
measure heat capacities per unit volume to about 10 parts per mil-
lion and, from these data, derive apparent (ϕ_C) and partial $(\overline{C}_p)$
molal heat capacities of each component. These $\overline{C}_p$ are a measure of
the intrinsic heat capacity of the component plus contributions
from the various interactions between the components. Being a second derivative of the chemical potential with respect to temperature, heat capacity is especially sensitive to changes in the structure of solutions and in particular to those resulting in energy
fluctuations in the system.

Organic molecules which are soluble in water usually have stan-
dard (infinite dilution) $\overline{C}_p^{\theta}$ which are significantly more positive
than the molar heat capacity C_p^o of the pure substances. With many
liquids which are completely miscible in water (dimethylformamide,
dioxane, acetone, etc.) $\overline{C}_p$ decreases in a fairly regular way from
$\overline{C}_p^{\theta}$ to C_p^o. However, with certain hydrophobic alcohols $(\underline{7})$, alkoxyethanols $(\underline{8})$ and amines $(\underline{9})$, $\overline{C}_p$ undergoes drastic changes in the
water-rich region, especially at lower temperature. Examples of
these changes are shown in Fig. 3 for *tert*-butanol, triethylamine
and 2-butoxyethanol in water. These systems show nearly a first or
higher order transition, and beyond 0.05 mole fraction the $\overline{C}_p$ are
constant and have the value of C_p^o, suggesting that in this concentration range the organic molecules locally seldom come into contact with water molecules. There is therefore some kind of micro-
phase separation occuring in these binary systems in a way which
could be analogous to micellization $(\underline{10})$. Continuing studies in
our laboratory seem to indicate that the addition of surface active
molecules stabilize further these pseudo-phases (transitions are
sharper and occur at lower concentrations). It is our contention
that the structures which exist in these binary systems are locally
not too different from those of many so-called microemulsions.

Microemulsions and Oil Recovery.

One of the interesting features of these binary solutions, and
of many microemulsions, is their tendency to unmix at higher temperature. For example triethylamine-water mixtures unmix into nearly
pure triethylamine and nearly pure water at 18.5°C; similarly 2-
butoxyethanol has a lower critical solution temperature at 49°C.
This phase separation suggests a new approach to the problem of
tertiary oil recovery. We can use such binary systems to dissolve
the oil at low temperature and then recover a good part of the oil
simply by raising the temperature some 20 to 30 degrees. This is
based on the assumption that, at high temperature, ternary systems
will also tend to separate into two phases, one of which would be
very rich in oil. This should be especially useful for the lighter

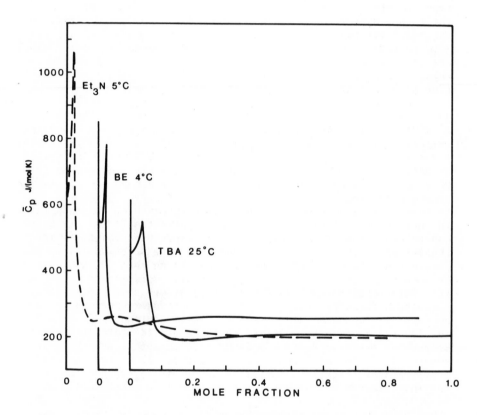

Figure 3.   Partial molal heat capacities of triethylamine, 2-butoxyethanol, and tert-butoxyethanol in water

oils.

Decane-2-Butoxyethanol-Water Phase Diagrams.

The phase diagram of some oil-solubilizer-water must be measured as a function of temperature in order to test the above approach. For this purpose decane (DEC) was chosen as a typical oil and 2-butoxyethanol (BE) as the solubilizer. We thought BE would be a good model solubilizer since the lower critical solution temperature for the BE-H_2O system is 49°C; this gives a good workable temperature range for our investigation.

In these experiments BE and DEC, from Baker Chemicals (Practical grade), were used as such. Sodium dodecyl sulfate (NaDDS), which was added to the solutions in some experiments, was from BDH (Specially pure). It was dried in a vacuum oven before use.

The phase diagrams were determined by the cloud point technique. At a constant temperature (Sodev temperature controler, ± 0.001°C) the BE-H_2O systems were titrated (2.5 ml Gilmont syringe) with DEC until a slight cloudiness appeared, corresponding to the formation of two or more phases. For temperature studies, the temperature of a known mixture of BE-DEC-H_2O was varied until a cloud point was observed. The temperature was read to ± 0.01°C with a pre-calibrated thermistor. In general this technique is fast and quite reproducible but difficulties are often encountered. In the water-rich region the solubility of DEC is low. At higher concentrations three phases often appear when excess DEC is added. The top phase contains mostly DEC and the lower one mostly water. The coexistence of three phases is typical of many microemulsions systems (11).

To determine the complete phase diagram of a ternary system as a function of temperature, at least a three-dimension diagram is necessary. Such diagrams are unfortunately quite difficult to visualize and it is often preferable to reduce the diagrams to two dimensions by keeping the concentration of some of the components constant. Some results for the BE-H_2O phase diagram as a function of temperature for fixed quantities of DEC are shown in Fig. 4. In this diagram the mole fraction of BE refers to the binary system BE-H_2O; *e.g.* X_{BE} = 0.4 means 0.4 mole BE in 0.6 moles H_2O. On the other hand X_{DEC} refers to the ternary systems; *e.g.* X_{DEC} = 0.2 means 0.2 moles of DEC in 0.8 moles of BE + H_2O.

On the left-hand side of Fig. 4 we have the normal phase diagram of the binary BE-H_2O system. The addition of DEC shifts the two phase equilibria to higher BE concentrations and to lower temperatures. If the addition of a third component is continued beyond the cloud point, eventually three distinct phases appear. Unfortunately the cloud point technique gives us the initial concentration or temperature where unmixing begins but is not suitable to distinguish between the coexistence of two phases and three phases. Also the three phase region depends quite critically on temperature.

At X_{DEC} higher than 0.1, two or more phases appear at high

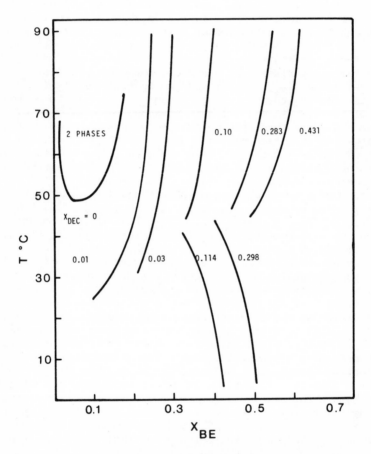

*Figure 4. Phase diagram of decane–2-butoxyethanol–water for a constant con-
centration of decane. The mole fraction of BE is expressed relative to water only
and the mole fraction of DEC relative to the BE–H₂O.*

temperatures and also at low temperatures. This probably suggests
that there exists also an upper critical temperature in the binary
system BE-H$_2$0 at low temperature. These lower and upper cloud
points were unfortunately not determined at exactly the same X_{DEC}.
Still we can readily see from Fig. 4 that for X_{DEC} near 0.1 and
X_{BE} = 0.4 the system will exist in two phases below 20°C, will be
completely miscible above this temperature and it will again separate out into two phases above 80°C. Again three phases seem to
appear when the cloud point is exceeded.

A similar phase diagram for a fixed X_{BE} and a variable X_{DEC} is
shown in Fig. 5. At low X_{DEC} the system is relatively simple. For
example, for X_{BE} = 0.45 and X_{DEC} = 0.225 two phases exist below
28°C (point a), a single phase between 28°C and 55°C (points b and
c) and again two phases above 55°C (point d). At high X_{DEC} the situation becomes very complex. Only one case is shown for X_{BE} =
0.70. At X_{DEC} = 0.755 a clear solution is observed below
20°C (point e), a cloudy mixture appears above 20°C but does not
unmix (point g), and two distinct phases are present above 35°C
(point h). We are possibly in a region where liquid crystals or
inverse micelles are formed.

The ternary phase diagram where the temperature is kept constant are shown in Fig. 6. Here X_{DEC} and X_{BE} are both expressed
relative to the ternary systems. In this diagram we show only the
cloud points corresponding to the initial unmixing. In the DEC-
rich region the diagram becomes too complex to fix unambiguously
the phase diagram. Data are shown for 25, 40 and 70°C. Again we
can see that for X_{BE} between 0.2 and 0.5 the solubility of DEC is
larger at 40°C than at 25 and 70°C.

The effect of adding a surfactant, (NaDDS), was also investigated. One such case only is shown in Fig. 6 where BE is replaced
by a 5:1 mixture of BE-NaDDS. The main effect of NaDDS is to increase the miscibility range of the oil in water. Various ratios
of BE-NaDDS were used and, as a first approximation, the change in
the phase diagram is directly proportional to the concentration of
NaDDS. The addition of a surfactant probably stabilizes the micro-
structures which were already present in the ternary system BE-DEC-
H$_2$0 and decreases the quantity of BE needed to solubilize DEC.
Therefore the presence of a surfactant is useful but not essential
to the stability of microemulsions.

Heat Capacity of Decane in Microemulsions.

Heat capacity measurements should be very useful in determining the local structure in microemulsions. A complete study will
involve keeping one component near infinite dilution and vary the
ratio of the other two. The standard $\bar{C}_p^0$ of the first component
will then inform us on the environment of the molecule. This
should be done for BE, DEC and H$_2$0 as the reference component.
Then in a second set of experiments the mole fraction of all components should be varied simultaneously. This is a long-term project and only preliminary results will be presented here for DEC as

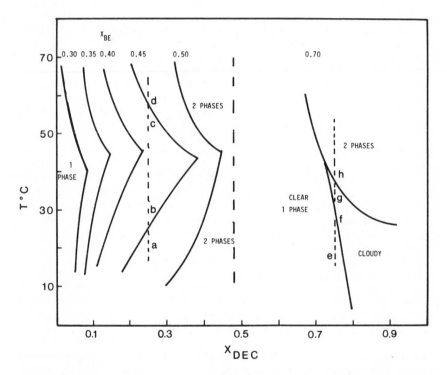

Figure 5. Phase diagram of decane–2-butoxyethanol–water for a constant concentration of 2-butoxyethanol. The mole fraction of BE is expressed relative to
water only and the mole fraction of DEC relative to the BE–H₂O mixture.

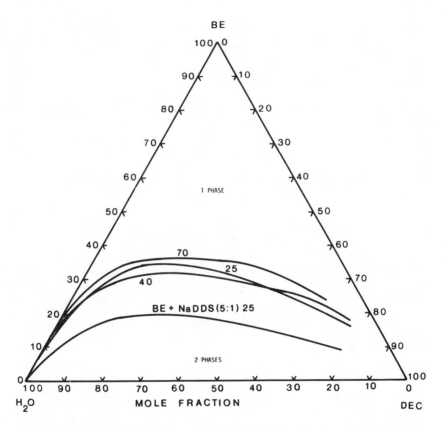

Figure 6. Phase diagram of decane–2-butoxyethanol–water for a constant temperature. All mole fractions are expressed relative to the other two components.

the reference component.

The apparent molal heat capacity ϕ_C of DEC was measured in BE-H_2O mixtures with a flow microcalorimeter $(6,7)$ following the usual procedure for ternary systems (12). In most experiments DEC was kept at very low molalities and X_{BE} was varied over a wide range (Table 1). In pure water and in very dilute BE solutions the solubility of DEC is too low for direct ϕ_C measurements. However $\bar{C}_p^\ominus$ of DEC in pure water can be estimated reasonably well through additivity rules $(\underline{13},\underline{14})$ and has the value 1060 J K^{-1} mol^{-1} at $25^\circ C$. On the other hand $\bar{C}_p^\ominus$ of DEC in pure BE is about 345 J K^{-1} mol^{-1}. Except for data points near X_{BE} = 0.05, which corresponds to the transition region, ϕ_C(DEC) does not vary much with the DEC or BE concentration and has a value close to $\bar{C}_p^\ominus$ of DEC in pure BE. This strongly suggests that DEC molecules are essentially in contact with BE only. This is again consistent with the hypothesis that the BE-H_2O system exists as microphases above 0.05 mole fractions.

TABLE I

HEAT CAPACITY OF DECANE IN 2-BUTOXYETHANOL-WATER MIXTURES AT $25^\circ C$.

m_{DEC} mol kg^{-1}	X_{BE}	ϕ_C(DEC) J K^{-1} mol^{-1}
0.0245	0.05	242.5
0.0329		291.6
0.0377		324.6
0.0476		356.8
0.0360	0.10	327.6
0.0630		331.0
0.0974		350.5
0.116		372.7
0.0268	0.20	341.5
0.0556		342.2
0.1113		339.3
0.2334	0.45	353.0
0.3303		352.0
0.9954	pure	345.3
1.5546		344.7

C_p^o of pure decane = 314.6 J K^{-1} mol^{-1}
$C_p^\ominus$ of decane in pure water = 1060 J K^{-1} mol^{-1}.

Conclusion.

The work presented here is essentially a preliminary study of the solubilization of oil in simple binary aqueous systems which can be used as model systems for microemulsions. Still, it was demonstrated that the use of phase separation to recover the oil from microemulsions is feasible. For this purpose BE is probably not the best solubilizer since the concentration range where the microemulsion can be destabilized by a change in temperature is too narrow and BE is not stable at high pH. Solubilizers like triethylamine and diethylmethylamine could be more interesting in this respect since the lower critical solution region is very flat. These systems are presently under study in our laboratory.

Acknowledgement.

JED is grateful to Imperial Oil and to the Quebec Ministry of Education for financial help. G.R. is also grateful for a France-Quebec exchange fellowship.

Abstract.

Microemulsions are very promissing for tertiary oil recovery in view of their capacity to dissolve large quantities of nonpolar molecules. Studies of the thermodynamics of organic aqueous mixtures have indicated that some polar organic liquids miscible with water, such as alcohols, amines and alkoxyethanols can exist in water as microphases or aggregates similar to micelles. These microphases can be further stabilized by the addition of third component such as a surfactant. As a model system the phase diagram of the ternary systems 2-butoxyethanol-decane-water was determined as a function of temperature. This ternary system unmixes at low and high temperatures. The effect of sodium dodecyl sulfate on the phase diagram was also investigated. The addition of a surfactant increases significantly the solubility of decane. Heat capacity measurements suggests that decane, when dissolved in a 2-butoxyethanol-water mixture, essentially comes in contact with the nonaqueous component only.

Literature cited.

1. Bansal, V.K., Shaw, D.O. "Microemulsions, Theory and Practice" edited by Prince, L.M., Academic Press, New York, 1977, chapter 7.
2. Prince, L.M. ref (1) chapter 1 and 2.
3. Prince, L.M. Ref (1) chapter 5.
4. Friberg, S. ref (1) chapter 6.
5. Picker, P., Leduc, P.A., Philip, P.R., Desnoyers, J.E., *J. Chem. Thermodyn.* (1971), 3, 631.
6. Desnoyers, J.E., de Visser, C., Perron, G., Picker, P., *J. Solution Chem.* (1976), 5, 605.
7. de Visser, C., Perron, G., Desnoyers, J.E., *Can. J. Chem.* (1977), 55, 856.
8. Roux, G., Perron, G., Desnoyers, J.E., *J. Solution Chem.* (in press).
9. Roux, G., Perron, G., de Visser, C., Desnoyers, J.E. (in preparation.

10. Roux, G., Perron, G., Desnoyers, J.E., J. Phys. Chem. (1978), 82, 966.
11. Friberg, S., Buraczewska, I., "Micellization, Solubilization and Microemulsions, edited by Millta, K.L., Plenum Press, New York, 1977, 791.
12. Avedikian, L., Perron, G., Desnoyers, J.E., J. Solution Chem. (1975), 4, 331.
13. Nichols, N., Sköld, R., Spink, C., Suurkuusk, J., Wadsö, I., J. Chem. Thermodyn. (1976), 8, 1081.
14. Perron, G., Desnoyers, J.E., Fluid Phase Equil. (in press).

RECEIVED July 18, 1978.

Desulfurization of Fossil Fuels

J. B. HYNE

Alberta Sulphur Research Ltd. and Department of Chemistry,
University of Calgary, Calgary, Alberta T2N 1N4

Since the industrial revolution mankind has been dependent
on fossil hydrocarbon fuel for a very large proportion of total
energy needs. Coal was the dominant form until the first half of
the twentieth century when fluid hydrocarbons began to make significant inroads due to easier recovery, handling and transportation. Many of the fluid forms also had lower sulphur content
than coal and were therefore more attractive from the environmental impact standpoint.

The low sulphur fuels, however, are limited in supply and
future energy needs indicate that, until alternate non-hydro-
carbon energy sources can be fully developed, there will be an
increasing reliance on heavier hydrocarbons, including coal,
heavy oils, bitumens, etc., all of which tend to be relatively
high in sulphur. Coupled with ever stricter environmental control regulations the importance of developing efficient technologies for fuel desulphurisation is clear.

This paper presents an examination of the overall question
of fossil fuel desulphurisation as it currently exists including
solid, liquid and gas phase hydrocarbon fuels and technologies
for their desulphurisation, in situ, in refining, during combustion and post combustion (see Figure 3).

Sulphur in Fossil Fuels

Sulphur occurs in most fossil hydrocarbon fuels in either or
both of the organic or inorganic forms. Pyrites (FeS_2) is the
most common of the inorganic forms while organosulphur compounds
range widely over the structural spectrum from thioethers and
mercaptans in the aliphatic series to thiophenes and complex aromatic structures where the sulphur atoms are an integral part of
the ring systems.

In many deposits, the amount of sulphur present in the hydro
carbon far exceeds the corresponding percentage of sulphur likely
to have been incorporated in the original biological material
from which the fossil fuel was formed. Thus the sulphur now pre-

sent in the hydrocarbon deposits was either concentrated during
the accumulation in the formation or has been incorporated from
adjacent inorganic sources.

Recent work by Wilson Orr (1) and others using stable isotope techniques has provided strong evidence that incorporation
from an inorganic source is the most probable explanation. Inorganic sulphur (sulphate) tends to be richer in the heavy sulphur isotope, S^{34}, compared with reference meteoritic sulphur
whereas organic sulphur is depleted in the heavy isotope. Using
typical values for δS^{34}, the relative enrichment or depletion
with respect to S^{34}, Orr showed that as much as 78% of the H_2S
associated with a hydrocarbon deposit and 34% of the organically
bound sulphur was derived from inorganic (formation rock) sulphate. A mechanism for this incorporation of sulphate sulphur
into organic structures is shown in Figure 1. Although hydrogen
sulphide acts as a crucial reaction intermediate undergoing redox
reaction with sulphate, the overall reaction involves the reduction of sulphate by hydrocarbon yielding sulphur (or H_2S) and
carbon dioxide. The elemental sulphur produced reacts readily
with hydrocarbons to yield organosulphur compounds typical of
those found in fuels. All of these reactions are promoted by
deep, high pressure, elevated temperature burial of the fossil
hydrocarbon. This effect is most dramatically seen in the high
concentrations of H_2S found in deep (10,000'+) natural gas wells
in many locations around the world.

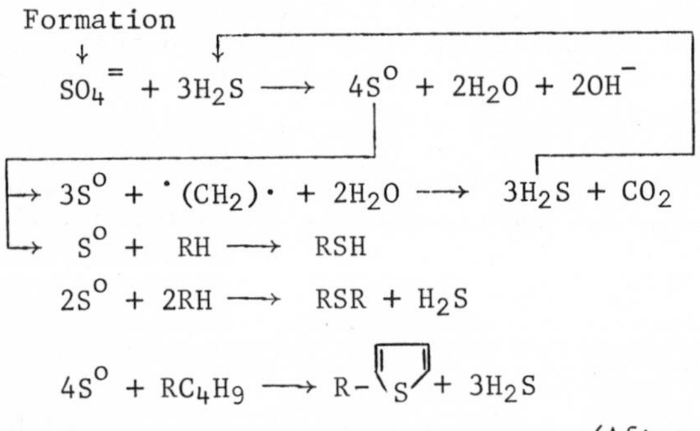

Formation

$$SO_4^= + 3H_2S \longrightarrow 4S^o + 2H_2O + 2OH^-$$

$$3S^o + \cdot(CH_2)\cdot + 2H_2O \longrightarrow 3H_2S + CO_2$$

$$S^o + RH \longrightarrow RSH$$

$$2S^o + 2RH \longrightarrow RSR + H_2S$$

$$4S^o + RC_4H_9 \longrightarrow R-\langle_S\rangle + 3H_2S$$

(After Wilson Orr, 1974)

Figure 1.   Organosulfur compounds from inorganic sulfate

Typical sulphur content values for various fossil hydrocarbon fuels are shown in Table I. The very high associated
sulphur content of many of the deep gas deposits is readily seen.
Fortunately, however, this sulphur content is present as relatively easily removed H_2S although the presence of such large
quantities at elevated temperatures and pressures can pose

serious corrosion and material failure problems. Both Canada and
France have built an extensive elemental sulphur production industry based on the recovered H_2S from sour natural gas.

Table I

Sulphur Content of Fossil Fuels

Gas Phase			$\%H_2S$
Natural Gas:	Alberta Foothills	(production)	54
	Mississippi	(production)	65
	Pyrenees	(production)	17
Liquid Phase		^{o}API	$\% S$
Crude Oil:	Venezuela, Buscan	10	5.6
	Mexico, Cretaceous	12	5.4
	Saudi Arabia, Manifa	28	3.0
	Saskatchewan, Weyburn	24	2.1
	Arkansas, Smackover	19	2.1
	Alberta, Swan Hills	37	0.8
	Louisiana, Delhi	44	0.08
	Alberta, Kaybob	47	0.04
Bitumen:	Athabasca Oil Sands	< 10	4.7
Solid Phase			
Coal:	Eastern North America		3 - 4
	Western North America		0.5
Lignite:	B.C., Sask., Ont.		2.0
Oil Shales:	Colorado		0.75

The sulphur content of crude oils varies markedly but is
generally related to the API gravity of the crude. The very
light crudes (API > 35) generally tend to be low in sulphur and
are therefore in considerable demand and command premium prices.
As discussed later, however, these low sulphur light crudes are
in relatively limited supply and the trend is clearly toward production of heavier, higher sulphur crudes. Venezuelan and Western Canadian heavy oils and bitumens represent the extreme end
of the API gravity spectrum and have correspondingly high sulphur
content. Since these deposits represent some of the world's largest reserves of fossil hydrocarbon fuel and will doubtless be
produced in ever increasing quantities the development of metho-
dology for desulphurisation is a matter of importance.
    While the sulphur values associated with gas, oil and bitu-
mens are either in the form of H_2S or organosulphur compounds
coals can have significant inorganic sulphur values. This introduces the possibility of desulphurisation by chemical methods
different from that required for removal of organic sulphur
values. Nonetheless, the desulphurisation problems associated

with the less tractable organosulphur compounds still exist with
coals and other solid phase forms. To date there has been only
very limited success in removing organosulphur values without
either gasification or combustion.

Figure 2 clearly shows that Canada's fossil fuel reserves
are heavily oriented toward coal, tar sand and heavy oils. These
huge reserves of some 800 billion barrels equivalent (c.f. current
Middle East oil reserves of some 400 billion barrels) leave little
doubt about Canada's long term fossil fuel self-sufficiency.
Converting these reserves into energy, however, will be dependent
on the development of new technologies, one of which must be effective containment of the inherent sulphur values in order to
minimise environmental impact. In 1967 it was estimated (2) that
of the 220,000,000 tons of sulphur emitted to the earth's atmosphere annually, 75,000,000 tons were man made and of that amount
50,000,000 tons resulted from coal combustion. If the extent to
which the industrialised world becomes dependent upon coal as an
energy source is to increase again, it is clear that removal of
sulphur values is essential to the protection of the environment.

In Situ Desulphurisation

Partial removal of the sulphur values in fossil hydrocarbon
fuels can be accomplished in place during the application of so-
called "in situ" recovery techniques. Perhaps the earliest of
these techniques to be explored was in situ coal gasification.
During the 1940's and 1950's, both the Russians and the British
carried out extensive field trials with in situ coal gasification
and demonstrated the practical feasibility of obtaining a low BTU
gas product from the process. More recently in situ coal gasification trials have been conducted in Alberta and several projects
in the Colorado oil shales have focussed on rubblising, and in
situ retorting of the shale to liberate the hydrocarbon. In situ
recovery techniques using fire, steam flooding, or a combination
of both, also appear to be one of the most practical routes for
fluidising the heavy bitumens of deeply buried oil sands thus
enabling their recovery by drilled wells rather than mining.

Perhaps the most significant advantage of such in situ techniques as far as desulphurisation is concerned, is that chemical
reaction temperatures and pressures can be reached that would require elaborate and expensive process equipment if attempted on
the surface. Thus chemical reaction regimes can be generated
"in situ" that would be less practical and economic if the fossil
fuel were recovered and subjected to traditional upgrading and
refining. These higher temperature and pressure conditions can
be of particular importance in desulphurisation reactions. Coker
units in surface equipment can probably simulate the hydrogen
transfer reactions that occur in "in situ" fire flooding yielding lower molecular weight hydrocarbons and hydrogen sulphide.
Thus in situ removal of the sulphur from its parent organosulphur

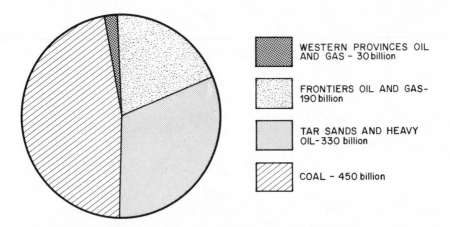

Figure 2.   Canada's potential recoverable reserves of fossil fuels (in barrels-of-oil equivalent).   Total: 1000 billion;   source: Federal Government Publication "An Energy Policy for Canada."

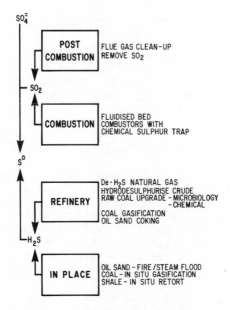

Figure 3.   Desulfurization of hydrocarbon fuels

compound is achieved and the more tractable gaseous H_2S form of sulphur can be removed and processed, eg.

$$R - S - CH_2 - R \xrightarrow[\text{thermal crack}]{\text{fire flood}} R - R + H_2S + C$$

$$\downarrow \qquad\qquad \downarrow O_2 \qquad \downarrow O_2$$

$$\text{desulphurised} \quad H_2O + S \quad CO_2$$
$$\text{hydrocarbon}$$

Less easily accomplished under readily accessible surface conditions, however, is the high temperature and pressure <u>hydrolysis</u> of organosulphur compounds, eg.

$$R - S - CH_2 - R + H_2O \xrightarrow[\text{press.}]{\text{high temp.}} 2RH + H_2S + CO$$
$$\text{steam flood}$$

In reactions of this kind not only is the sulphur removed from the hydrocarbon but the hydrogen of the water (steam) becomes intimately involved in the overall process. Indeed the reaction is the combination of the thermal cracking shown in the fire flood example and the water gas reaction between the produced coke and steam.

$$R - S - CH_2 - R \longrightarrow R - R + H_2S + C$$

$$C + H_2O \longrightarrow H_2 + CO$$

$$\overline{R - S - CH_2 - R + H_2O \longrightarrow 2RH + H_2S + CO}$$

Recent work (<u>3</u>) with model organosulphur compounds has shown that at temperatures above 350°C and pressures in excess of 100 atm (1500 psig), the hydrolytic desulphurisation reaction occurs readily with thioethers, mercaptans and other non-thiophenic types of organosulphur compounds. Thiophene itself is more resistant to this type of reaction but desulphurisation is significant in the 450 - 500°C range. More complex fused ring sulphur containing aromatic structures, can, however, be more reactive.

Clearly there are considerable advantages to removing sulphur from its organic skeleton "in situ" and recovering it at the surface in the more readily handled H_2S form. Although the generation of H_2S in such in situ recovery processes may introduce special requirements in materials selection due to corrosion and embrittlement problems the sulphur converted into H_2S is sulphur that does not consume hydrogen in hydrodesulphurisation refinery steps or appear in residual coke in coker upgrading of heavy oil or bitumens.

Much remains to be learned about the reactions of organosulphur compounds at the elevated temperatures and pressures that can be readily achieved in in situ recovery processes. The use of the natural formation as the chemical reactor permits the attainment of reaction conditions that have previously been out

of the range of practical interest of the organic chemist. The
hydrolysis of organic compounds has been a common area of study
in the past but few have ventured into the domain of hydrolysis
above 100°C. Similarly, the high temperatures attainable in in
situ retorting and combustion (gasification, fire flooding) will
require a better understanding of molecular rearrangement, cracking, disproportionation and similar hydrocarbon reactions in a
reaction temperature and pressure regime that has received relatively little attention.

Refinery Desulphurisation

Desulphurisation of hydrocarbon fuels has traditionally been
carried out primarily as part of the refining and upgrading process. Accordingly by far the most advanced and best understood
chemistry and technology is to be found in this area. Prior to
the advent of major concern for environmental impact of fossil
fuel combustion products relatively little was done to desulphurise hydrocarbon fuels (principally coal) prior to combustion and
past effects of large scale consumption of high sulphur coals can
still be seen in major industrialised areas around the world.

The first major attempt at precombustion desulphurisation
was in the coal gas industry and a number of efficient and effective techniques for removal of H_2S, COS, CS_2, mercaptans and
other volatile sulphur containing products of the gasification
process were developed. Many of these techniques found application in the subsequent development of sour natural gas processing where large volumes of hydrogen sulphide had to be removed from the hydrocarbon component.

Oil   Perhaps the most intensive chemical research effort in
hydrocarbon desulphurisation, however, has been directed toward
the removal of sulphur values from crude oils. It is not appropriate in a general desulphurisation overview of this type to attempt to do other than place the whole area of traditional crude
oil desulphurisation in perspective in relationship to other
hydrocarbon desulphurisation. By far the commonest technique has
been hydrodesulphurisation requiring the generation of large
quantities of molecular hydrogen for the catalytic extraction of
organosulphur values from the oil as hydrogen sulphide. Many
proprietry hydrodesulphurisation techniques have been developed
by the major refiners each utilising some particular advantageous
feature of a catalyst, temperature or pressure reaction conditions
(4). Perhaps the most elegant aspect of much of this prior research has been in the area of catalyst development where not only
efficiency but resistance to poisoning by the other constituents
of the crude oil feed have represented major challenges.

Sulphur content of crude oils has always been regarded as a
discounting factor and as environmental restrictions on sulphur
emissions have been tightened, the need to desulphurise refinery

feedstocks has increased. In addition, the ever growing demand
for hydrocarbon fuels and the limited supplies in industrialised
countries has brought into the feedstock stream heavier crudes
with higher sulphur content. Only 36% of some 13 MM bbl/d. of
crude run to U.S. refineries in 1973 was above 0.5% sulphur, but
this had jumped to 46% of a larger 15 MM bbl/d. in January 1978
(5). The trend is also illustrated by a comparison of the hydro
processing (including hydrodesulphurisation) capacities in both
the U.S. and Western Europe between 1967 and 1977. U.S. hydroprocessing capacity increased from 30.5% to 43.5% during this
period while the European figures show an even more dramatic jump
from 12% in 1967 to 31% of crude capacity in 1977 (6).
     This trend to heavier and higher sulphur content feedstocks
is likely to continue. As noted earlier, the heavy oils and
bitumens of both Canada and Venezuela, are rich in sulphur and
will undoubtedly be desulphurised in ever increasing amounts as
supplies of the lighter, lower sulphur content feedstocks dwindle.
     It is always dangerous to conclude that little remains to be
done in a particular research field. More than likely, having so
concluded, a significant new breakthrough is immediately announced! Considering the relatively undeveloped nature of the many
other aspects of hydrocarbon desulphurisation, however, it would
seem that further progress in the area of catalytic hydrodesulphurisation of liquid hydrocarbon feedstocks is likely to be less
dramatic than in the past. This does not mean, however, that
there is no need for continued effort in improving hydrodesulphurisation processes particularly with respect to the techniques
for handling the heavy crude compositions and feedstocks with
high thiophenic type organosulphur content.

Gas   Desulphurisation of natural gas consists primarily of removal of hydrogen sulphide. Thus the sulphur content of the gaseous form of fossil hydrocarbon fuel is already in the form to
which it is normally converted in other precombustion desulphurisation techniques. Despite this advantage, however, the desulphurisation of sour natural gas can have problems associated with
the sheer volume of hydrogen sulphide involved. H_2S content of
15 - 25% is commonplace in sour gas fields around the world but
gas with 50 - 65% is already being processed in both Canadian and
U.S. fields. Canadian sour gas processing plants handle some
20,000 tons per day of hydrogen sulphide extracted from the natural gas production of the foothills gasfields of Alberta and
British Columbia. As the drill penetrates into deeper, hotter
$(200^{\circ}C)$ and high pressure (1500 atm) formations in search for
more gas and oil the hydrogen sulphide content of the gas generally rises. Deep sour gas wells in the Southern U.S. have shown
65% H_2S content and wells with as high as 90% H_2S have been completed in the Canadian Rockies. Processes for handling these
high concentrations of H_2S in such large volumes must clearly
differ from those previously available for handling the off-gases

from refinery hydrodesulphurisation or coal gas manufacture where
H_2S levels are significantly lower.

Alkanolamines have been widely used for removal of H_2S from
sour natural gas. Monoethanolamine (MEA) was typically used to
react with the acidic H_2S to form the alkanolammonium sulphide.
This salt could be readily removed

$$2CH_2 - CH_2 + H_2S \rightleftharpoons \left(\begin{array}{cc} CH_2 - CH_2 \\ | \quad\quad | \\ OH \quad\quad NH_3 \end{array} \right)^+_2 S^=$$

$$\begin{array}{cc} | \quad\quad | \\ OH \quad\quad NH_2 \end{array}$$

from the hydrocarbon gas stream by the aqueous solution and the
separated H_2S regenerated by heating with live steam. Diethanol-
amine (DEA) and diisopropanolamine (DIPA), however, have replaced
the simpler MEA because of irreversible reaction of MEA with the
small amounts of COS and CS_2 that also occur in sour natural gas.

Hydrogen sulphide, however, is not the only acid gas present
in sour gas. It is normally accompanied by significant quantities of carbon dioxide which also reacts with the amine scrubbing
solution. This is beneficial from the standpoint of reducing the
non-combustible content of the natural gas thus increasing its
fuel value, but it adds to the load that the amine must carry and
competes with the H_2S. Differential absorption of H_2S and CO_2
therefore becomes a factor of considerable importance in processing sour natural gas and combinations of physical and chemical
absorption have been developed such as in the Sulfinol Process
where diisopropanolamine (chemical reactant) and tetramethylene
sulphone (physical solvent) are used together. The ability of
most systems to discriminate between H_2S and CO_2 under equilibrium conditions is usually not great but the rate at which these
two acid gas components react with basic agents can be significantly different. Thus it may be possible to design a process
where the desulphurisation of sour natural gas can be achieved
more efficiently by the selective removal of H_2S using kinetic
rather than equilibrium discrimination vis a vis carbon dioxide.

Coal   Precombustion desulphurisation of coal has been largely
associated with coal gasification. Removal of inorganic sulphur,
mainly pyrites, and to some extent the more frequently occurring
organic sulphur has been possible by so-called "coal cleaning"
involving physical or chemical leaching methods. Present technology has been recently reviewed by Davis (7). Physical separation of sulphur values is limited to the inorganically bound
sulphur and has little if any effect on the 40 - 70% organosulphur content. The differential density of pyrite and coal is
the primary property utilised in cyclone, concentration tables
and froth flotation techniques. Some attention has been given
to magnetic techniques for pyrite separation.

Chemical leaching of pyrite has been accomplished with
aqueous ferric sulphate under pressure. The overall reaction is

$$FeS_2 + Fe_2(SO_4)_3 \xrightarrow{H^+} 3FeSO_4 + 2S \downarrow$$

oxidation

a redox system whereby the ferric iron is reduced to ferrous
(sulphate) and the sulphide of the pyrite is oxidised to elemental sulphur. The latter is vaporised from the coal and the reduced ferrous sulphate can be re-oxidised to the ferric state by
oxygenation.

Removal of organic sulphur from the solid coal phase normally
requires oxidation. Treatment of pulverised coal with alkaline
sodium carbonate and aeration at pressures up to 10 atm has been
reported (8) to remove almost all inorganic and some 30% of organic sulphur. The sulphur values undergo oxidation in the process and

$$Coal - S + Na_2CO_3(alk) \xrightarrow[press.]{O_2} Coal + Na_2SO_4$$

$$CaSO_4 \leftarrow$$

$$CaCO_3 \xrightarrow{heat} CaO + CO_2$$

appear as water soluble sodium sulphate. The sodium carbonate
can be regenerated by treatment of the sulphate with lime and
CO_2 from a limestone kiln. The extracted sulphur thus appears
finally as calcium sulphate waste.

It has been said that a "bug" can be found to do any chemical task - and do it better! While the all embracing scope of
such a claim may be somewhat exaggerated, there is little doubt
that modern microbiology is playing an ever increasing role in
industrial process design. Coal desulphurisation by bacteria has
recently been improved by the combined use of thiobacillus ferrooxidans and thiooxidans (9). The ferrooxidans have long been
known as the cause of acidic mine drainage by oxidising pyrite to
sulphur acid. In combination with the sulphur consuming thiooxidans and adequate aeration of the coal slurry being treated,
much faster desulphurisation is possible. Sulphur content of coal
has been reduced from 5% to 2% in nine days with promise of reduction to 1% levels.

Desulphurisation of the hydrocarbon values of coal as an
integral part of the gasification process remains the most active
and attractive precombustion technique. Despite the number of
specific gasification processes (10) (Lurgi, Koppers-Totzek, Bi-
Gas, Hygas, Cogas, Synthane, etc.), the essential features of
coal gasification can be summarised in the diagrammatic form
shown in Figure 4. The inorganic and organic sulphur forms in
the coal are largely converted into the more readily handled
hydrogen sulphide form in the initial gasifier step. The H_2S
content of the low thermal content gas from the gasifier is
normally quite low (0.4% by volume) and high capacity techniques

such as employed in sweetening high H_2S content sour natural gas
are not required.   Both physical and chemical absorption techniques as outlined in Table II are used (11).   The Selexol and
Rectisol processes employ alcohols or their ether derivatives as
physical solvents for the acid gases (CO_2 and H_2S) while the
Sulfinol Process, as described previously combines the physical
solvent properties of tetramethylene sulphone with the chemical
reactivity of an alkanolamine.   The Benfield process, otherwise
known as the "hot pot" chemically absorbs the acid gases in aqueous potassium carbonate solution and the Jefferson DGA system
again utilises chemical reaction of the CO_2 and H_2S with amine to
yield the alkanolammonium salt.   Regeneration of all these scrubbing systems and recovery of the H_2S and CO_2 rich acid gas stream
can be readily effected by heating.

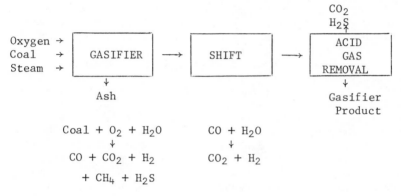

Figure 4.   Generalized coal gasification sequence

Table II

Acid Gas (H_2S, CO_2) Absorbers

Process	Licensor	Absorber	
Selexol	Allied Chemical	Polyethyleneglycol Ether	PHYSICAL
Rectisol	Lurgi	Methanol	
Benfield	Benfield	Aqueous Potassium Carbonate	CHEMICAL
DGA	Jefferson Chem.	Diglycolamine	
Sulfinol	Shell	Tetramethylenesulphone + Diisopropanolamine	

Because of the high concentrations of CO_2 compared with H_2S found in coal gasifier product streams the relative concentrations of H_2S and CO_2 and the absorption capacity of the various scrubbing systems for the two components can be a crucial factor in selecting a process. Also of importance is the point in the overall gasification sequence at which the acid gases and in particular H_2S is removed. Several types of catalyst can be employed in the downstream "shift" reaction where the low thermal content gasifier product can be upgraded by reaction of some of the CO content with water to increase the hydrogen concentration. Some of these shift catalysts are sensitive to H_2S and in such cases desulphurisation of the gasifier product must be accomplished ahead of the shift reactor.

If the gasifier product stream is intended for downstream use as the feedstock for further upgrading such as methanation, methanol or Fischer Tropsch synthesis, very thorough desulphurisation is essential since the catalysts in these upgrading processes are highly sensitive to sulphur poisoning. The methanation catalysts normally cannot tolerate more than 0.05 ppm of sulphur in the feedstock. In addition to H_2S sulphur values in the gasifier product it may contain COS, CS_2, mercaptans and thiophenes. These are normally removed by activated carbon or zinc oxide filters ahead of the sensitive synthesis catalyst beds.

Oil Sand Bitumen   Desulphurisation of oil sands bitumen with 4.5% organosulphur content may, in the in situ fire or steam flood processes, be partially accomplished underground. But for the mining operations currently in place and in the planning stage the desulphurisation process must occur during the refining or upgrading stages. Hydroprocessing of extracted bitumen will result in the sulphur values appearing as H_2S in the off-gases and this sulphur form can be removed by any of the several absorption methods discussed previously.

However, a significant proportion of the initial 4.5% sulphur in the bitumen is retained in the coke production from the fluid or delayed cokers used to redistribute the hydrogen content of the bitumen. In the present GCOS delayed coking process for upgrading Athabasca oil sands bitumen some 20% of the carbon of the original bitumen is produced as coke containing almost 6% sulphur. The larger Syncrude operation will convert some 8% of initial bitumen to coke containing 8 - 9% sulphur. The fuel values of these huge coke productions can only be realised if combustion is possible within the limits of environmental tolerance. Thus there is a continuing strong interest in methods of desulphurising coke that will avoid the necessity for installation of flue gas desulphurisation after combustion.

Techniques similar to that discussed earlier for the desulphurisation of coal have been used. Treatment of sulphur containing coke with lime can yield calcium sulphide which can be converted back to lime with generation of H_2S by hydrolysis.

Hydrodesulphurisation of coke, however, may be economically feasible (12). While this technique requires a source of hydrogen this, as will be discussed shortly, may be possible without a large net consumption of the hydrogen. Of particular interest is the observation that pre-treatment of the coke with caustic (13) and pre-oxidation (14) can significantly improve hydrodesulphurisation yields.

<u>Hydrogen from H_2S</u> Throughout the discussion of refinery desulphurisation the common sulphur containing end product has been H_2S. After extraction from the main feedstock stream by an appropriate physical or chemical absorption method the H_2S concentration is normally high enough to be fed to a Claus reactor where the H_2S is converted to elemental sulphur and water.

$$2H_2S + O_2 \longrightarrow 2H_2O + 0.25S_8$$

Thus the hydrogen generated for use in the upstream hydrodesulphurisation step emerges from the system as water – environmentally acceptable but very much a "once through" chemical utilisation of this valuable reagent.

Considerable attention has been given recently to the direct recovery of the hydrogen from hydrogen sulphide rather than conversion to water. Catalytic thermal cracking of H_2S is possible (15,16) and improved catalysts permitting thermal decomposition at lower temperatures are being investigated.

$$H_2S \underset{\substack{\text{catalyst} \\ \text{heat}}}{\rightleftharpoons} H_2 \underset{\text{removal}}{\downarrow} + S \downarrow$$

An important feature of any successful process based on catalytic thermal decomposition is that the system must be kept in an "upset equilibrium" condition since the equilibrium concentration of hydrogen in the presence of sulphur and H_2S is low at readily accessible temperatures. By use of a flow system and separation of the reaction products, however, the yield of hydrogen can be markedly improved by utilising the relative kinetics of the forward and reverse reactions.

Recovery of the hydrogen by chemical reaction is also feasible (17). Many lower sulphides of metals will react readily with H_2S to yield hydrogen and the higher sulphide of the metal. The higher sulphides are usually unstable at elevated temperatures and regeneration of the lower sulphide can be effected with recovery of elemental sulphur.

$$MS_x + H_2S \rightleftharpoons MS_{x+1} + H_2$$
$$\uparrow \underline{\quad -S \quad}|$$
$$\text{Heat}$$

Such processes for the recovery of hydrogen from H_2S are energy consuming compared with the Claus treatment of the product H_2S which is highly exothermic. The overall economics of a successful process may, however, be such as to make hydrogen recovery from H_2S more attractive than continual hydrogen generation for a "once through" hydrodesulphurisation role.

Combustion Desulphurisation

As costs of precombustion hydrodesulphurisation and post combustion flue gas clean-up have escalated and as environmental regulations have further limited the sulphur dioxide emission rates, there has been a growing interest in technology designed to effect fuel desulphurisation during the combustion process. Desulphurisation during fluidised bed combustion of coal has been a leading technique in these developments.

The principle of fluidised bed combustion with simultaneous desulphurisation is based on the thermal decomposition of limestone carbonates to yield oxides which then react with the sulphur oxide products of combustion of both inorganic and organic sulphur compounds in the hydrocarbon fuel.

$$CaCO_3 \longrightarrow CaO + CO_2$$

$$Coal - S \xrightarrow{O_2} CO_2 + H_2O + SO_2$$

$$CaO + SO_2 \xrightarrow{O_2} CaSO_4$$

ash

Combustion temperatures are high enough to ensure carbonate decomposition and full oxidation of the sulphur values to the sulphate form.

The ratio of limestone to fuel required for effective desulphurisation depends upon both the mineral and sulphur content of the coal. It may be possible to burn low sulphur (<0.8%) western coal directly with 90% sulphur retention thus avoiding the necessity for flue gas desulphurisation and still meeting environmental SO_2 emission regulations (18). Fluidyne (19) have indicated that a 3.65% Illinois coal with 0.3 lbs. of dolomite added per pound of coal in a fluidised bed combustor can reduce SO_2 emission to less than 1.2 lbs. of SO_2 per million BTU thermal output thus meeting U.S. EPA limits for large plants.

Coals with relatively high carbonate or metal oxide content may require significantly less added limestone since the natural mineral content will also act as a sulphur trap in the fluidised bed. "Self desulphurisation" of this type can range as high as 60% with some coals (20). Although temperatures in the combustion bed must be high enough to decompose the carbonates SO_2 absorption has also been shown to decrease as temperature rises. Thus an optimum combustion temperature range must be sought that permits adequate carbonate decomposition and efficient SO_2 absorption. This is normally in the range 800 to 900°C (20). These lower combustion temperatures also help to minimise the formation of NO_x. Only the nitrogen content of the fuel will lead to some NO_x formation at these temperatures.

Post Combustion Desulphurisation

Desulphurisation of hydrocarbon fuels prior to combustion has been seen to be primarily achieved by reducing the inherent sulphur values to hydrogen sulphide. In post combustion desulphurisation the sulphur values are almost exclusively in the oxidised SO_2 form. While this is hardly surprising in view of the oxidative nature of the fuel combustion process, it does mean that essentially different chemistry is involved and the nature of the oxidant - air - introduces large volumes of inert diluent - nitrogen.

The dilution, or low concentration factor is a major concern in any post combustion desulphurisation technique. Sulphur concentrations are not particularly high in the raw fuel prior to combustion, usually less than 6%. Thus when the oxidant air brings with it a further 80% nitrogen diluent the sulphur values in the combustion effluent are even further diluted. While from the standpoint of atmospheric dispersal through high stacks this dilution is advantageous, it obviously makes the problem of recapturing and removing sulphur from the effluent that much more difficult. Thus, in addition to purely chemical problems post combustion desulphurisation faces the entropy challenge of removing a low concentration component with very high efficiency.

Perhaps the most difficult task in reviewing and comparing the multitude of processes that have been, and are being proposed for desulphurisation of effluent gas streams is that of finding a suitable framework upon which to organize the data. Without such a structure, the very volume of information is simply unmanageable.

The first necessity is to reduce the number of individual processes considered by selecting typical examples of various types. Many of the more than fifty known processes are modifications of a general type of system, and considerable simplification can be achieved by selecting illustrative examples. An effort has been made to identify and reference related processes that are not discussed in detail.

A major subdivision of processes is possible on the basis of the principal phase in which the reactions take place. This clearly cannot be a clear-cut division since many processes have a number of steps which involve, in turn, the gas phase, the liquid or solution phase, and the solid phase. By and large, however, the principal characteristic step of each process is associated with either the gas phase or solution, and on this basis a significant subdivision can be made.

Processes occurring principally in the gas phase are considered in Table III. These processes can be generally of the
oxidative type leading to sulphuric acid or sulphates with sulphur in the +6 oxidation state, or they can be overall reduction
processes yielding elemental sulphur in the zero oxidation state.
These characteristics permit a further binary classification of
the gas phase processes.

The chemistry of the TVA Dry Limestone process (21) is essentially identical to that discussed previously in connection
with desulphurisation during fluidised bed combustion. Better
temperature control can be effected, however, since the SO_2
absorption is now separated from the combustion process. The
end product slag ($CaSO_4$) requires suitable disposal and although
this is not seen as a serious problem, some concern has been
voiced regarding calcium ion poisoning of surface waters if excessive leaching of slag disposal sites occur. The Swedish
Bahco Process (22) is similar but uses hydrated lime slurry.

Manganese dioxide can be used to absorb the initially low
concentration SO_2 to produce the sulphate in a Mitsubishi Process
(23). In this case the absorbing phase is itself the oxidising
agent. Regeneration with ammonia and air simultaneously produces
ammonium sulphate which can be directly marketed as a fertiliser,
thus the calcium ion problem of the dry limestone process is replaced by a plant nutrient ion - NH_4^+.

The Active Magnesia Process (24) can be viewed as the gas
phase equivalent of the Wellman-Lord (to be discussed below) in
that it serves primarily to concentrate the SO_2 values in the
combustion effluent to a suitable concentration for feed to an
oxidative process yielding sulphuric acid. The absorption temperature, however, is quite low compared with typical combustion
effluent temperatures.

Processes 4, 5 and 6 are all essentially one step oxidations
of SO_2 to SO_3 and hence sulphuric acid. The first pair are modified versions of the traditional Contact and Chamber processes
for sulphuric acid manufacture, with the principal change being
in their ability to accept dilute SO_2 gas streams as the feedstock. The use of Activated Carbon as an air oxidation catalyst
has clearly received international attention, with success or
failure depending to a large extent on subtle modifications in
catalyst preparation and catalyst presentation to the reactant
gases. Virtually every type of catalyst bed configuration has
been explored.

Processes 7 to 9 in Table III have the common feature of
producing sulphur as the end product. They are therefore overall reduction processes. The Cat-Redox and CO/SO_2 systems are
both direct conversions to elemental sulphur. The latter has a
particular advantage of being able to accept effluent gas streams
at temperatures of $750°F$ or higher thus avoiding the need for any
cooling of combustion flue gases before clean-up. The Alkalized
Alumina Process, however, involves concentration and catalytic

hydrogenation of the sulphur components to hydrogen sulphide for
subsequent oxidation to elemental sulphur in a Claus Process.
It is therefore related to such processes as the Alkazid (25)
which involve preliminary hydrogen reduction to H_2S before
Claus treatment.

Solution processes for removal of SO_2 from effluent gas
streams normally require lower absorption temperatures than do
gas phase techniques. Thus, of the four listed in Table IV only
the Molten Salt Method has the capability of accepting high
temperature flue gas without cooling. In all but one of the
cases, elemental sulphur is the end product and in the single
instance of the Ammoniacal Solution Process, sulphur is a co-
product with ammonium sulphate. This process has been exten-
sively examined and developed in a number of countries, and is
chemically interesting because of the unusual redox reaction
that is suspected to take place between the products of air oxidation of SO_2 absorbed in ammonia solution. Both products,
sulphur and ammonium sulphate, are normally saleable commodities.

The use of molten salts in process chemistry is, relatively
speaking, a recent development. The high temperatures involved
require special handling techniques but, in the case of molten
carbonates and thiocyanates, the temperatures involved are
wholly compatible with the temperatures characteristic of the
effluent gas streams to be desulphurised. Both the simple
molten carbonate (Atomics) and carbonate/thiocyanate (Garrett)
processes involve reduction of a metal sulphite to the corresponding metal sulphide, but the reducing conditions differ
through use of reformed natural gas in one case and coke roasting in the other. In both cases, the resulting H_2S is fed to
a Claus plant for re-oxidation to elemental sulphur. Again, as
was seen in the Alkalized Alumina case reduction to sulphur is
not direct, but via the -2 oxidation state of H_2S and subsequent re-oxidation to elemental sulphur.

Although at first sight, the Citrate Process may not appear
to be in any way related to the traditional Claus, it is in fact
an H_2S/SO_2 redox reaction in solution with the activating bauxite,
carbon, or metal salt type catalyst replaced by a citrate complex
with SO_2. The chemistry of the process is clearly interesting
and of some importance but for the purposes of this review it is
sufficient to draw the analogy indicated above. The Citrate
Process is yet another reduction process that may require the
ancillary generation of H_2S from natural gas and product sulphur
if the effluent gas stream is solely SO_2 as far as sulphur content is concerned.

The Wellman-Lord Process is not, in itself, a conversion
method, but rather a solution phase technique for concentrating
a dilute SO_2 effluent stream to provide a suitably rich feed for
Claus redox conversion. When coupled with the Claus Process, it
constitutes an overall desulphurisation system which involves
all three phases: gas, liquid solution, and solid crystalline.

Table III

Gas Phase Desulphurisation

No.	Process (Developer) (Reference)	Process Chemistry
	OXIDATION TO SULPHATE	

1 Dry Limestone (TVA) (21)

$$SO_2 + Air + CaO \longrightarrow CaSO_4$$
$$CaCO_3 \xrightarrow[\text{heat}]{} CO_2$$

2 Manganese Dioxide (Mitsubishi) (23)

$$SO_2 \xrightarrow[MnO_2]{} MnSO_4 \xrightarrow[\text{Air}]{NH_4OH} (NH_4)_2SO_4 + H_2O$$

Regeneration

3 Active Magnesia (Showa Hatsuden) (Chemico) (24)

$$SO_2 \xrightarrow[MgO]{200 - 300^0 F} MgSO_3 \xrightarrow{1400^0 F} 15\% \ SO_2 \quad Gas$$

Regeneration

Contact Process

$$H_2SO_4$$

4 Modified Contact (Monsanto-Penelec) (Tokyo Tech.) (26)

$$Air + SO_2 \xrightarrow[V_2O_5]{900^0 F} SO_3 \xrightarrow[H_2O]{} H_2SO_4$$
$$\xrightarrow[2 \ NH_4OH]{} (NH_4)_2SO_4 + H_2O$$

5 Modified Chamber (Tyco Labs. Mass) (27)

$$SO_2 + H_2O + NO_2 \longrightarrow H_2SO_4 + NO$$
$$1/2 \ O_2$$

6 Activated Carbon (Sulfacid-Lurgi) (Hitachi, Tokyo) (Reinluft, W. Germany) (Westvaco Corp. U.S.) (28, 29, 30, 31)

$$SO_2 \xrightarrow[\text{Active Carbon}]{Air, H_2O} H_2SO_4$$

| | **REDUCTION TO SULPHUR** | |

7 Catalytic Redox (Princeton Chem. Res.) (32)

$$SO_2 + H_2S \xrightarrow[\text{Catalyst}]{250 - 350^0 F} H_2O + S \longrightarrow$$
$$CO_2 \leftarrow \qquad \leftarrow CH_4$$
Catalyst, High. Temp.

8 CO/SO$_2$ Redox (Chevron Research) (Univ. Mass.) (33, 34)

$$SO_2 + CO \longrightarrow 2 \ CO_2 + S$$
$$1000^0 F, Cu \ on \ Al_2O_3 \qquad NO \ present \ converted \ to \ N_2$$

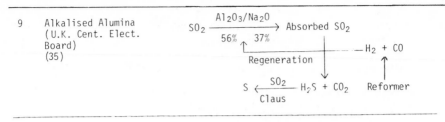

9   Alkalised Alumina
    (U.K. Cent. Elect.
    Board)
    (35)

$SO_2 \xrightarrow[\substack{56\% \quad 37\%}]{Al_2O_3/Na_2O}$ Absorbed SO_2

Regeneration

$S \xleftarrow[\text{Claus}]{SO_2} H_2S + CO_2$ Reformer

$— H_2 + CO$

Table IV

Solution Phase Desulphurisation

No.	Process (Developer) (Reference)	Process Chemistry

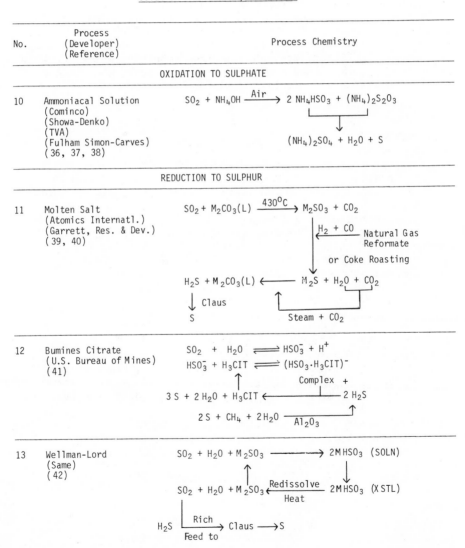

OXIDATION TO SULPHATE

10   Ammoniacal Solution
     (Cominco)
     (Showa-Denko)
     (TVA)
     (Fulham Simon-Carves)
     (36, 37, 38)

$SO_2 + NH_4OH \xrightarrow{Air} 2 NH_4HSO_3 + (NH_4)_2S_2O_3$

$(NH_4)_2SO_4 + H_2O + S$

REDUCTION TO SULPHUR

11   Molten Salt
     (Atomics Internatl.)
     (Garrett, Res. & Dev.)
     (39, 40)

$SO_2 + M_2CO_3(L) \xrightarrow{430^oC} M_2SO_3 + CO_2$

$H_2 + CO$ — Natural Gas Reformate

or Coke Roasting

$H_2S + M_2CO_3(L) \longleftarrow M_2S + H_2O + CO_2$

↓ Claus

S   Steam + CO_2

12   Bumines Citrate
     (U.S. Bureau of Mines)
     (41)

$SO_2 + H_2O \rightleftharpoons HSO_3^- + H^+$

$HSO_3^- + H_3CIT \rightleftharpoons (HSO_3 \cdot H_3CIT)^-$

Complex +

$3 S + 2 H_2O + H_3CIT \longleftarrow$ — $2 H_2S$

$2 S + CH_4 + 2 H_2O \xrightarrow{Al_2O_3}$

13   Wellman-Lord
     (Same)
     (42)

$SO_2 + H_2O + M_2SO_3 \longrightarrow 2 MHSO_3$ (SOLN)

$SO_2 + H_2O + M_2SO_3 \xleftarrow[\text{Heat}]{\text{Redissolve}} 2 MHSO_3$ (XSTL)

$H_2S \xrightarrow[\text{Feed to}]{\text{Rich}}$ Claus $\longrightarrow$ S

The heart of the Wellman-Lord concentration step involves the separation of SO_2 as the crystalline metal bisulphite and subsequent thermal decomposition of the redissolved material.

Conclusion

A very substantial technology for the desulphurisation of fossil hydrocarbon fuels has been developed over the past half century. Although much of the recent advance has been in desulphurisation of liquid and gas phase hydrocarbons useful improvements have been made in previous technologies for the removal of sulphur values from the solid phase fuel-coal. The clear need to bring on stream the heavier and higher sulphur containing hydrocarbons as an energy source for the future presents fresh challenges to the chemist and chemical engineer to develop new and more efficient desulphurisation techniques which can help bring to the marketplace economically priced fuels that will not have a negative environmental impact.

Literature Cited

1.   Orr, Wilson L., Bull. Am. Assoc. Pet. Geol., (1974) 58 (11), 2295-2318.
2.   Stanford Research Institute Project PR-6755 (1967).
3.   Macdonald, D.D., Greidanus, J.W., Hyne, J.B., "Hydrothermal (Water) Reactions of Athabasca Bitumen Organosulphur Model Compounds and Asphaltene", The Oil Sands of Canada-Venezuela 1977, CIM Special Volume 17.
4.   Beaton, W.I. in "Chemical Reactions as a Means of Separation - Sulfur Removal", Chem. Proc. and Eng. Series, II (1977) Ed. B.L. Crynes, Marcel Dekker Inc., New York, Chapter 1 and references therein.
5.   Oil & Gas J. Newsletter, March 13 (1978).
6.   International Petroleum Encyclopedia 1977, The Petroleum Publishing Co., Tulsa (1977), 295.
7.   Davis, J.C., Chem. Eng., March 1 (1976) 70-74.
8.   Chem. Eng., January 16 (1978) 73.
9.   Coal Age, January (1978) 149.
10. Linden, H.R., Bodle, W.W., Lee, B.S., Vyas, K.C., Annual Review of Energy, 1 (1976) 77.
11. Christensen, K.G., Stupin, W.J., Hydrocarbon Proc., Feb. (1978) 125-130.
12. U.S. Patent 2,721,169 (1955).
13. U.S. Patent 3,009,781 (1957).
14. Parmar, B.S., Tollefson, E.L., Can. J. Chem. Eng., 55 (1977) 185.
15. Kotera, Y., Int. J. Hydrogen Energy, 1 (1976) 219; U.S. Patent 3,962,409 June 8 (1976).

16. Chivers, T., Carle, D., Hyne, J.B., "The Catalytic Thermal Decomposition of Hydrogen Sulphide", Proc. 5th Can. Sym. on Catalysis, Oct. 26-27 (1977) Calgary, Pub. CIC.
17. Weiner, J.G., Legette, C.W., U.S. Patent 2,979,384 (1961).
18. Kolisnyk, Z., German-Canadian Sym. on Coal Refining, Edmonton, Alberta, April 20-21 (1978).
19. Fluidyne Eng. Corp., Minneapolis, Minnesota 55422.
20. Schilling, H.D., Schreckenberg, H., Wied, E., German-Canadian Sym. on Coal Refining, Edmonton, Alberta, April 20-21 (1978).
21. Tennessee Valley Authority, Knoxville, Tenn., (1968).
22. Chem. and Eng. News, Sept. 6 (1971) 19.
23. Bovier, R.F., Proc. Am. Power Con., (1964) 26, 138.
24. Chem. and Eng. News, Aug. 30 (1971) 18.
25. Leuhddemann, R., Noddes, G., Schwarz, H., Oil & Gas J., (1959) 57, (32) 100.
26. Stites, Horlacher, Bachoffer and Bartman, Proc. Am. Soc. Mech. Engineers, 68-WA/APC-2, Winter Meeting, New York, December 1968.
27. Sauder, Wallitt, Rissmann and Keilin, Tyco Labs., Proc. Sym. on Air Pollution Control Tech., 65th National Meeting, AIChE, Cleveland, May (1969).
28. Staub-Reinhalt Luft (1968) 28, 6.
29. Staub-Reinhalt Luft (1968) 28, 1.
30. Chem. and Eng. News, Sept. 8 (1969) 48.
31. Oilweek, July 5 (1971) 9.
32. Chem. and Eng. News, Sept. 2 & 9 (1968) 10 & 22 res.
33. J. Air Pollution Control Assoc., (1967) 17 (12) 796.
34. Chem. and Eng. News, Dec. 14 (1970) 54.
35. Bienstock, D., Field, J.H., Myers, J.G., U.S. Bureau of Mines Report of Investigations 7021, Wa. D.C. (1967).
36. King, R.A., Ind. Eng. Chem., (1950) 42, 2241.
37. Simon-Carves Ltd., British Patent 525,883, Sept. 6 (1940).
38. Chem. and Eng. News, Sept. 8 (1969) 48.
39. "Sulphur", The British Sulphur Corp. Ltd., London (1969) (81) March-April.
40. Chem. and Eng. News, Apr. 12 (1971) 65.
41. Chem. and Eng. News, June 14 (1971) 31.
42. Poeter, B.H., Paper at CNGPA Meeting, Calgary, Nov. 26 (1971)

RECEIVED September 25, 1978.

6

The Relation of Surfactant Properties to the Extraction of Bitumen from Athabasca Tar Sand by a Solvent–Aqueous-Surfactant Process

DONALD F. GERSON, J. E. ZAJIC, and M. D. OUCHI

Biochemical Engineering, The University of Western Ontario,
London, Ontario, Canada

The Athabasca tar sands are an enormous petroleum resource, but this supply of hydrocarbons will be difficult to utilize without considerable economic and energetic expenditure. Present technology involves mining and hot-water extraction, future technology will center around *in situ* extraction methods. In both cases the energetic cost of hot water or steam production sharply reduces overall net recovery (up to 25–35%) since petroleum energy must be put into the extraction process. A paramount consideration is, therefore, the study of methods and strategies which reduce the energy requirements of the extraction, demulsification and upgrading processes, thus increasing yield and decreasing ultimate economic costs.

Athabasca tar sand is a 3–component system (sand–bitumen–water) dominated by surface effects. It is estimated that there is between 0.5 and 1.5 square meters of bitumen–sand interfacial area per gram of tar sand. The process of separating the bitumen from the sand thus clearly involves overcoming or modifying the adhesional energies between the bitumen and this very large mineral surface. The hot water system currently in use effects a reduction in this interfacial energy by the direct saponification of complex carboxylic and resin acids present in the bitumen to produce surfactants which reduce interfacial tensions and facilitate separation (1,21). Reduction of the interfacial tensions of the tar sand is crucial to both the mining and *in situ* methods of recovery, and yet systematic study and the experimental modification of these interfacial tensions is virtually absent from the literature.

The most important step in any aqueous extraction scheme is facilitation of the transfer of mineral particles from the bituminous matrix to the aqueous phase (Fig. 1). This process appears to be more favorable in the Athabasca tar sands than in other tar sands due to a postulated, but unobserved, film of absorbed water present on the mineral particles. The transfer process may be visualized in 2 stages: 1) the transition from complete immersion in the bulk bituminous phase to partial contact with both phases, and 2) the transition from partial contact with both phases to complete immersion in the aqueous phase (Fig. 1a,b).

If the net free energy of this process, ΔG_{net}, is negative the process will occur provided activation energies are overcome (15). The free energies of each stage can be either positive or

$$\Delta G_{net} = \Delta G_a + \Delta G_b \qquad (1)$$

negative, For extraction with water alone at room temperature, ΔG_{net} appears to be negative, the process is spontaneous but incomplete. Addition of mechanical energy appears to aid in the reduction of ΔG_a, while the addition of surfactants to the aqueous phase appears to aid in the reduction of ΔG_b. Both of these free energies depend on the solid-liquid free energies, which, until the work discussed by Neumann (17), have been difficult to measure. Appropriate and extensive application of these methods to the problem of tar sand extraction has only begun, but the results are promising. A review of the relevant literature may be found in the work of Zajic and Gerson (6,7,8,24) and McCaffery and Mungan (14).

Methods

Surface Tension Measurements. Surface, interfacial and adhesion tensions were measured with a Fisher Autotensiomat, a de Nuoy surface tensionmeter modified to accept a sensitive strain gauge. Surface and interfacial tensions were measured with a platinum loop, and adhesion tension was measured against small slabs of pyrophyllite (an hydrous aluminum silicate mineral, $AlSi_2O_5(OH)$, chosen for its resemblance to montmorillonite clay (11) present in the tar sand and responsible for many of the separation difficulties encountered). Measurement of immergent and emmergent adhesion tensions allows calculation of both advancing and receding contact angles, and can give an estimate of the surface tension of the solid being studied (17).

Tar Sand Extractions. Tar sand extraction was measured by 2 methods developed by Gerson, et al. (6,8). One method, the low-shear aqueous-surfactant extraction, utilizes low shear mixing provided by a rotary, planar microbiological shaker, to effect separation which is highly dependent on surfactant type. Yields by this method are low but it is extremely useful in the preliminary screening of surfactants. The second method utilizes a small paddle mill rotating a low speed (50–100 rpm) to make a slurry of tar sand, kerosene and water which is then diluted and stirred vigorously to yield 3 phases: a surface bituminous phase, an intermediate clear aqueous phase, and a dense mixture of sand, bitumen and water. The surface phase is virtually free of mineral matter and the bitumen contained in it is used to calculate percent bitumen recovery. This method is detailed by Gerson, et al. (8). This procedure attempts to simulate a cold water extraction process which utilizes both solvent and surfactant.

Surfactant Properties. Hydrophyllic-lipophyllic balance (HLB) was calculated from equation 2 (3). Molecular weights and chemical compositions were obtained from the literature or from the manufacturers. The partition coefficients and HLB values were not

$$HLB = 0.36 \ln (C_w/C_o) + 7 = 0.36 \ln K_p + 7 \qquad (2)$$

where: C_w = equilibrium aqueous concentration
C_o = equilibrium oil concentration
K_p = water-oil partition coefficient

available for the Makon surfactants (nonyl-phenyl-poly-(ethoxy)-ethanol surfac-

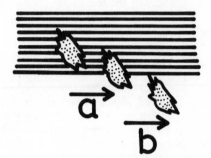

*Figure 1. Transfer of mineral particle
from bituminous matrix to the aqueous
phase*

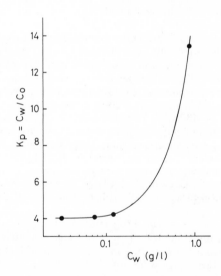

*Figure 2. Effect of surfactant concentration in the aqueous phase (Cw) on the
partition coefficient (Kp) of Makon-14*

tants of variable polymer chain length, produced by Stepan Chemical Co., and a gift of C. Tennant Co.). These were determined spectrophotometrically at 273nm (aqueous) or 277nm (hexadecane). Equilibration was effected by 2 hours of vigorous agitation followed by 30 min. centrifugation at 15,000xg. Triton-X surfactants were a gift of Rohm and Haas, and Tween surfactants were a gift of Atlas Chemicals.

Results

Partition Coefficients of nonyl-phenyl-poly-(ethoxy)-ethanol (NPE) Surfactants. The solubility of surfactants in water and hydrophobic solvents is well documented (11,12,22), but only a few attempts at measuring partition coefficients between immiscible liquids have been reported (2,4,9,10). Partition coefficients of surfactants are of theoretical interest because of their relation to observed surfactant properties such as emulsification, wetting and detergency. Partition coefficients (K_p) may be also of considerable practical value for predicting surfactant recovery and recycling in industrial processes. For example, in the cold water extraction of tar sand, an effective surfactant with a high K_p could be efficiently recycled in the process water and would not follow the bitumen into the upgrading stream.

Partition coefficients of surfactants have been reported to remain constant below the critical micelle concentration (CMC), and to increase with concentration above the CMC (2,9,10). The effect of surfactant concentration in the aqueous phase (C_w) on K_p was investigated with Makon 14 (14 mol% ethylene oxide, NPE_{14}), the results are given in Fig. 2. These data indicate a CMC of about 0.lg/l, or 12 μM, in close agreement with the value obtained by surface tension measurements (our data and ref. 22). In subsequent determinations of K_p, C_w was just below the CMC to minimize the effects of micellization (15,23).

Partition coefficients of NPE surfactants were determined for hexadecane-water mixtures (Table 1). Similar results were obtained by Crook et al. (2) for octyl analogs in an isooctane-water system. However, Log K_p is not a linear function of the mol% ethylene oxide in the surfactant as predicted by Eq. 2. Nonlinearity in the octyl analogs was due to polydispersity in the polymer chain length (2), and similar effects presumably operate here.

Tar Sand Extraction with NPE Surfactants. Aqueous-surfactant extractions of bitumen from tar sand utilizing the NPE series of surfactants (0.02%w/v) reveal that maximum release of bitumen occurs at an HLB of 6.7−7.0 (8−10 mol% ethylene oxide). Included in this set of data are the results for Triton-X surfactants (octyl analogs) having 5,11, and 12 mol% ethylene oxide. Several processes occur in the room temperature (25±3 C) extraction of tar sand by this low shear technique, these are described in Fig. 3. As the tar sand is slowly deformed by gentle shaking, new sand and clay grains are exposed to the surface (stage 1a, above), these are then released into the aqueous phase (stage 1b). This ablation results in an enrichment of the bitumen content of the tar sand sample. The bitumen concentration in the residual tar sand sample is plotted as B/RTS in Fig. 3, and is maximal at 20 mol% ethylene oxide (HLB=8). The sand particles have some adherent bitumen (% total bitumen in sand, Fig. 3), in practice this was the major loss in a pilot plant project utilizing a cold water process (Gerson, Cooper, Zajic, unpublished results). Losses of bitumen on released sand minimizes

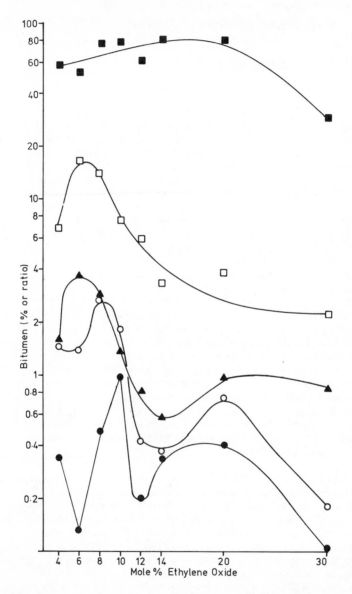

Figure 3.   Aqueous–surfactant tar sand extraction with NPE surfactants (0.02%
w/v). (■) Bitumen concentration in residual tar sand (B/RTS); (□) percent of
total bitumen in sand fraction; (△) bitumen concentration in sand fraction (B/S);
(○) percent of total bitumen in surface fraction; (●) composite index of extrac-
tion efficiency.

TABLE I

**Water-Hexadecane Partition Coefficients of NPE's[a]
and Calculated Values of HLB[e]**

n [b]	HLB from this data	$K_p = C_w/C_o$ [c]
4	5.83	0.039
6	6.33	0.157
8	6.73	0.477
10	7.01	1.04
12	7.39	2.96
14	7.62	5.52
20	7.97	14.7[d]
30	8.47	60.0[d]

[a] Original concentration = 0.2 g/l
[b] Average mole ratio of ethylene oxide
[c] Average of four samples, unless specified otherwise
[d] Average of two samples
[e] Calculated from HLB = 0.36 ln K_p + 7

$$C_9H_{19} - \underset{\bigcirc}{} - (OCH_2CH_2)_n - OH$$

n = average mole ratio of ethylene oxide

at the high end of the HLB scale studied here. In addition, a proportion of the bitumen floats to the surface of the aqueous phase (%total bitumen on surface, Fig. 3), this curve has 2 maxima, one near the low end of the range and one near the high end of the HLB range. A composite index reflecting the net beneficial effects of the surfactant peaks at an HLB of 7 (see 6,7,8,24).

The paddle mill was used to study the effect of surfactant type on a solvent-aqueous-surfactant extraction scheme for the recovery of bitumen from Athabasca tar sand. In the experiments of Figures 4,5 and 6, bitumen recovered from the surface phases was measured as a function of the mole fraction of ethylene oxide in the surfactant and as a function of the extraction step in which the surfactant was added. The results are reported as the % of the total bitumen present in the surface fraction. The amount of surfactant used was that required to give a final aqueous concentration of 0.02% (w/v), but in different sets of experiments the surfactant was added at various stages in the process.

Fig. 4 gives the results for experiments in which all of the water used contained surfactant at a concentration of 0.02% (w/v). Under these operating conditions, maximal extraction occurred with surfactant containing 15–20 mol% ethylene oxide. This corresponds to an HLB of 7.6–8.0. The maximum value obtained was 22% of the total bitumen in the surface fraction (Makon 20). Fig. 5 gives the results for experiements in which all the surfactant was added in a small volume of water at the beginning of the process, resulting in a much higher initial

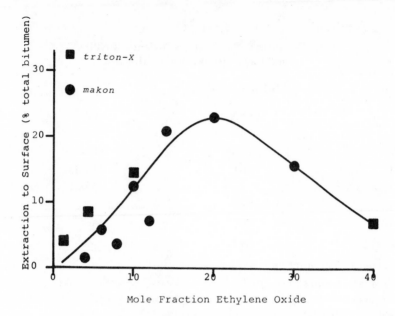

Figure 4. Solvent–aqueous-surfactant extraction of tar sand with Triton and Makon surfactants. Surfactant added to total aqueous phase.

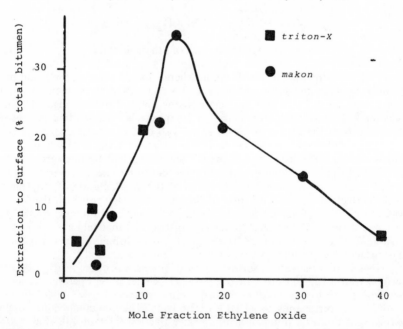

Figure 5. Solvent–aqueous-surfactant extraction of tar sand with Triton and Makon surfactants. Surfactant added to initial aqueous phase.

concentration of surfactant. Greater extraction was obtained with the optimal surfactant under these conditions than was obtained with the previous operating conditions: with Makon 14 (14 mol% ethylene oxide, HLB = 7.6) 34% of the total bitumen was present in the surface fraction. Also, under these conditions, the extraction is more sensitive to the HLB of the surfactant. Fig. 6 gives the results for experiments in which the surfactant was added to the kerosene, and distilled water was used throughout. More scatter is present in this set of data and is most probably due to solubility problems for the surfactants having greater than 10 mol% ethylene oxide. A sharp optimum was obtained for Makon 12 (HLB = 7.4), with 34% of the total bitumen present in the surface fraction. Since the HLB scale greatly compresses the range of partition coefficients, the data of Fig. 6 are replotted as a function of K_p in Fig. 7 for the Makon surfactants. This analysis reveals a steady increase in extraction up to a cut-off level at which recovery drops precipitously.

These data demonstrate that more hydrophobic surfactants (lower HLB) are more effective as the situation in which they are initially exposed to the tar sand becomes less aqueous. In addition, this trend correlates with increased overall extraction.

Adhesion Tensions and Tar Sand Extraction with Tween Surfactants. Measurement of the adhesion tension (τ) allows the determination of the wettability of a given solid by a given liquid or surfactant solution. Measurements of adhesion tension between both bitumen or clay surfaces and various surfactant solutions is thus highly relevant to a study of the effects of surfactants in the separation of bitumen from Athabasca tar sand.

Measurements have been made of the adhesion tension between 0.02% (w/v) solutions of Tween 20, 40, 60 and 80 and platinum, pyrophyllite, mica, bitumen saturated pyrophyllite and hydrated pyrophyllite. Adhesion tensions against these solids are given in Fig. 8 as a function of the HLB values for these surfactants, in general there is no simple correlation between these properties, indicating that there is a relatively strong and dominating interaction between the solid and the surfactant. Absorption may be an important aspect of these results, but will, of course, also be part of the extraction process. Fig. 9 gives measured values of adhesion tension (AT) as a function of measured values of spreading tension (aqueous solution vs. 20% bitumen in kerosene). The adhesion tension against mica is independent of spreading tension and that of platinum decreased with increasing spreading tension. Adhesion tensions against pyrophyllite depend on the pre-treatment of the surface, but in both cases have a maximum at a spreading tension of about −6 dynes/cm. Fig. 9 also gives the relation between HLB and spreading tension for these surfactant solutions.

Fig. 10 relates the composite extraction index (see above) obtained in the low-shear aqueous test system for these Tween surfactants, and adhesion tensions measured against various solids. Adhesion tensions against platinum and bitumen saturated pyrophyllite are irregularly related to tar sand extraction, while the adhesion tension against a fresh pyrophyllite surface is linearly (inversely) related to tar sand extraction. This is the first linear correlation between a measurable property of a surfactant solution and tar sand extraction which we have been able to obtain, and there appears to be no such finding in the literature. Fig. 11 gives the relations between extraction of bitumen with the paddle mill, solvent-aqueous-surfactant extraction and adhesion tensions measured against platinum, bitumen saturated pyrophyllite and hydrated (48 hours in water) pyrophyllite.

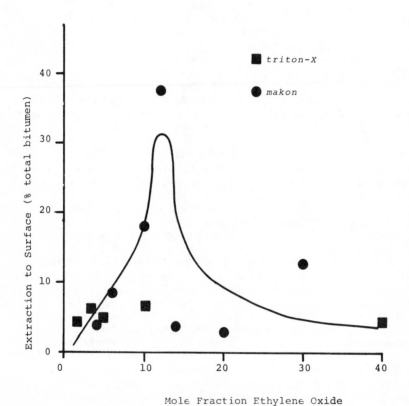

Figure 6.   Solvent–aqueous-surfactant extraction of tar sand with Triton and
Makon surfactants.   Surfactant added to solvent (kerosene).

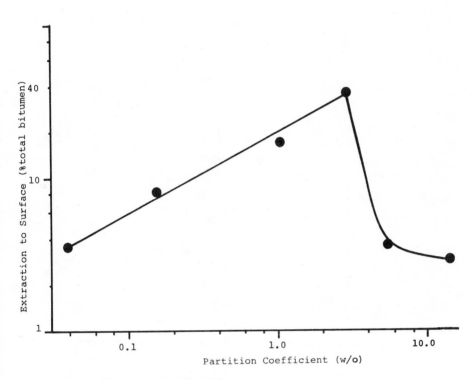

*Figure 7.   Solvent–aqueous-surfactant extraction of tar sand vs. partition coeffi-
cient for Makon surfactants added to solvent (kerosene)*

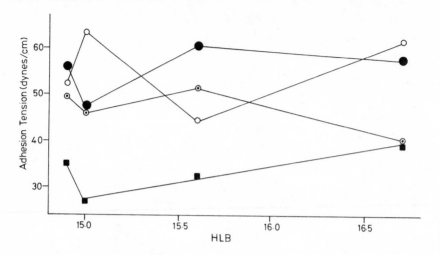

Figure 8. *Adhesion tension vs. HLB, Tween surfactants. (●) Fresh surface of pyrophyllite; (○) hydrated pyrophyllite; (⊙) bitumen-saturated pyrophyllite; (■) platinum.*

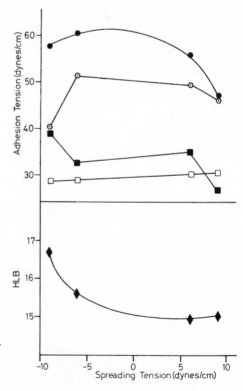

Figure 9. Adhesion tension and HLB vs. spreading tension, Tween surfactants. Symbols as in Figure 8; (□) mica; (♦) spreading tension against a 20% (w/v) solution of bitumen in kerosene.

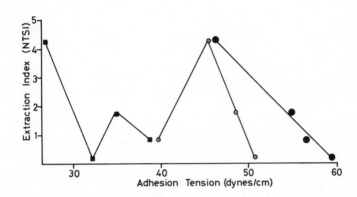

Figure 10. Aqueous-surfactant extraction of tar sand (composite extraction index) vs. adhesion tension against various solids for Tween surfactants (0.02% w/v). Symbols as in Figure 8.

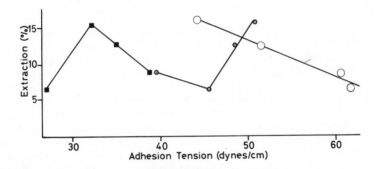

Figure 11. Solvent–aqueous-surfactant extraction of tar sand vs. adhesion tension against various solids for Tween surfactants (0.02% w/v). Symbols as in Figure 8.

A linear correlation is obtained between bitumen extraction with the paddle mill and the adhesion tension against water saturated pyrophyllite. That the degree of water saturation of the pyrophyllite is important in explaining the difference between the 2 extraction processes indicates that it will be necessary to study each process in terms of the relevant adhesion tensions. These results demonstrate that adhesion tension is the most important parameter found to date in determining the degree of separation in the presence of surfactants. Measurements of adhesion tension between surfactant solutions and minerals similar to those found in tar sand may be of considerable value in studies of surfactant utility in both aqueous-surfactant, solvent-aqueous-surfactant and *in situ* extraction processes. In addition, if appropriate model situations can be developed, measurements of adhesion tension may be useful in upgrading bitumen-water-clay emulsions obtained by a variety of *in situ* and heavy oil recovery processes.

Conclusions

These studies clearly indicate that investigation of the tar sand from the point of view of surface physical chemistry can be of practical benefit in optimizing and understanding processes for bitumen recovery. A very considerable literature exists which discusses the effects of surface and interfacial tensions in liquid-solid systems (eg. 1,5,19,20). Some literature exists on the surface chemistry which is relevant to tar sand separation (eg. 1,14). However, adequate solution of both energetic and economic problems associated with both the extraction of tar sand and the breaking of emulsions produced during extraction will depend on detailed knowledge on the interfacial physics relevant to a given process.

Abstract

The extraction of bitumen from Athabasca tar sand by a combined solvent-aqueous-surfactant (cold-water) method at 25±3 C and neutral pH offers many advantages over the present extraction technology which utilizes high pH at 80 C. Foremost of these are reduced energy requirements and reduced clay-sludge disposal problems. Any cold-water extraction scheme avoiding these problems with surfactants depends on surfactant selection criteria. To date, few systematic studies of the relations between measurable surfactant properties and the extraction of bitumen from tar sand have been performed. In this work, the relations between bitumen extraction in a solvent-aqueous-surfactant test system and the properties of the surfactant solution have been measured for a series of octyl and nonyl-phenyl-poly-(ethosy)-ethanol surfactants. The surfactant properties measured included surface, interfacial and spreading tensions, partition coefficients and HLB (hydrophillic-lipophyllic balance). Optimal bitumen extraction with these surfactants depended not only on surfactant properties but also on the details of the extraction process. When used in aqueous solution, optimal extraction occurred at spreading tensions of 10−20 dynes/cm, and at HLB values of 7. Adhesion tensions between pyrophyllite and aqueous solutions of Tween surfactants are linearly correlated to bitumen recovery by both aqueous-surfactant and solvent-aqueous-surfactant extraction methods. This appears to be the first linear correlation between surfactant properties and the extraction of bitumen from tar sand which has been observed.

Literature Cited

1. Bowman, C.W., 7th World Pet. Cong. Proc. (1967), 3:583.
2. Crook, E.H., D.B. Foddyce and G.F. Trebbi, J. Coll. Sci. (1965), 20:191.
3. Davies, J.T., Proc. 2nd Int. Cong. Surf. Act. (1957), 1:426.
4. Davies, J.T. and E.K. Rideal, Interfacial Phenomena, Academic Press, New York, 1963.
5. Davis, B.W., J. Coll. Sci. (1975), 52:150.
6. Gerson, D.F., J.E. Zajic and A. Margaritis, Microbial Separation of Bitumen from Athabasca Tar Sand, Univ. Western Ont. Biochem. Engr. Research Reports Vol. 1, 1976.
7. Gerson, D.F. and J.E. Zajic, in: Redford, D. (ed) The Oil Sands of Canada-Venezuela, (1978), CIMM 17:705.
8. Gerson, D.F., J.E. Zajic, D.G. Cooper and M.D. Ouchi, Microbial Separation of Bitumen from Athabasca Tar Sand Part 2, Univ. Western Ont. Biochem. Engr. Research Reports, Vol. 3, 1978.
9. Greiger, P.F. and C.A. Kraus, J. Am. Chem. Soc. (1949), 71:1455.
10. Greenwald, H.L., E.B. Kice and J. Kelly, J. Anal. Chem., (1961), 33:465.
11. Hurlbut, C.S., Dana's Manual of Minerology, 17th ed., Willey, 1965).
12. Jungermann, E. (ed), Cationic Surfactants, Dekker, New York, 1970.
13. Linfield, W.M. (ed), Anionic Surfactants, Dekker, New York, 1976.
14. McCaffery, F.G. and N. Mungan, J. Can. Pet. Tech. (1970), 21:185.
15. Mizrahi, J. and E. Barnea, Prog. Heat Mass Trans. (1971), 6:717.
16. Mukerjee, P. and K. Mysels, Critical Micelle Concentrations of Aqueous Surfactant Systems, NBS (USA) RDS 36, 1971.
17. Neumann, A.W., Adv. Coll. Int. Sci. (1974), 4:105.
18. Neumann, A.W., R.J. Good, C.J. Hope and M. Sijpal, J. Coll. Int. Sci., (1974), 49:291.
19. Rapacchietta, A.V., A. W. Neumann and S.N. Omenje, J. Coll. Int. Sci. (1977), 59:541.
20. Rapacchietta, A.V. and A.W. Neumann, J. Coll. Int. Sci. (1977), 59:555.
21. Sanford, E., 3rd AOSTRA Universities Seminar, Banff, 1978.
22. Schick, M.J., S.M. Atlas, F.R. Eirich, J. Phys. Chem. (1962), 66:1326.
23. Schick, M.J. (ed), Nonionic Surfactants, Dekker, New York, 1967.
24. Zajic, J.E. and D.F. Gerson, in: Strausz, O. (ed) Oil Sands and Oil Shale, Springer-Verlag, 1978.

RECEIVED September 25, 1978.

7

Laboratory Simulation of In-Situ Coal Gasification

M. GREENFELD

Fuel Sciences Division, Alberta Research Council, Edmonton, Alberta, Canada T6G 2C2

To provide laboratory support for the Alberta Research
Council's underground coal gasification field test program (1),
and means for exploring novel operating procedures before taking
them into the field, a gasification simulator has been developed.
This facility has been designed to reproduce the conditions of a
coal seam undergoing gasification, but eliminates peripheral
matters (e.g., water incursion) and consequently allows detailed
study of reaction kinetics and related aspects (e.g., cavity
formation, sweep efficiency and heat losses). In contrast to
previous laboratory work, which generally centered on small coal
blocks or fixed bed reactors, and commonly sought to define
limiting conditions which could be correlated with mathematical
models, the ARC simulator employs a 1 x 1 x 2m (3' x 3' x 6')
block which retains most of the essential features of an undis-
turbed coal seam.

Simulator Design

The general arrangement of the simulator assembly is shown
in Figure 1 and is comprised of a flow control system, a reactor,
a data acquisition system and over-ride controls that ensure
safety in operations.

The flow control system provides facilities for injecting air,
steam, oxygen or other fluids into the reactor, and is governed by
d/p cell transmitters and pneumatically-activated valves which
permit automatic or manual regulation of flow rates and pressures.

All reagent fluids are first mixed at predetermined pressure
and temperature in a mixing tank before they are sent through an
inlet gas control valve into an auxiliary preheater and the
reactor. The latter is a refractory-lined, rectangular sheet-
metal vessel; and after a coal block has been closely fitted into
the reactor, one or more 3/4" - 2" diameter horizontal holes, which
serve as initial reaction channels, and appropriately placed
vertical injection and production holes are drilled into it.
An electric heater element, inserted to the bottom of the

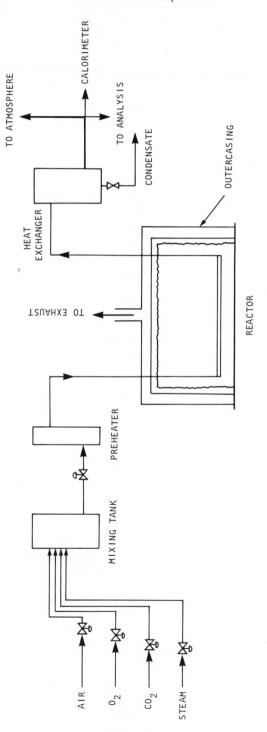

Figure 1. UCG simulator-simplified flow diagram

injection hole, is used to ignite the coal, and subsequent gasification then simulates underground gasification by the so-called
stream method.

Spaces between the coal block and the reactor walls are
filled with refractory cement to prevent accumulation of combustible gases, and leaks are exhausted from the outer casing.
Temperatures are measured with 1/8" Type K SS sheathed thermocouples, which are cemented into the block at predetermined
locations.

Product gas is cooled, freed of particulate matter by passage
through a filter, and then sent through a heat exchanger where
water and tars are condensed. Flow rates are metered as the
cleaned gas leaves the heat exchanger. Gas compositions are
determined with two programmable gas-chromatographs and a gas-
partitioner (which serves as backup unit for hydrogen determinations), and heat values are continuously monitored with a
modified gas calorimeter. All instrumentation for gas analysis
is calibrated with certified cylinder gases.

The data acquisition network is schematically shown in
Figure 2 and makes use of two computer programs. One - 'Therm'
- acquires temperature-data from thermocouples via an Autodata 8
scanner, stores these data, and displays them graphically on a
line printer for every fifth sample group (see Figure 7). The
second program - 'Balance' - first prompts and acquires basic
data relating to the progress of gasification and to product gas
compositions, and then requests from the control terminal information about a number of other parameters (e.g., flow rates)
which it prints out as a report on mass-, heat- and energy-
balances. All data are stored in disc files from which they can
be retrieved for further processing after completion of a test
run.

The safety system was designed with regard for the particular
laboratory space and the building as a whole.

Experimental

The progress of gasification in a previously drilled reaction
channel is schematically illustrated in Figure 3. This shows the
principal reaction zones and indicates pyrolysis immediately
ahead of gasification; and since the nature of pyrolysis products
is strongly temperature-dependent (2,3), the advance of each
reaction zone to any point in time can be quite closely determined
by correlating temperature profiles in the coal with the chemical
composition of the reactions products. From studies of coal combustion, it is estimated that 90 percent of the total burning time
of a particle relates to burning of the devolatilized (char)
particle (4). However, in underground gasification, where a
large excess of coal exists, the rate of coal consumption may be
expected to be zero order and is, in fact, found experimentally to
be independent of air flow rates, but strongly dependent on tem-

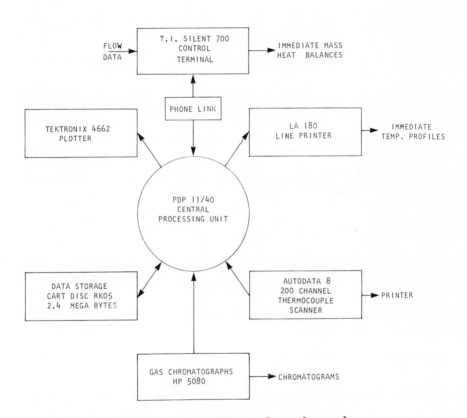

Figure 2. Data acquisition and control network

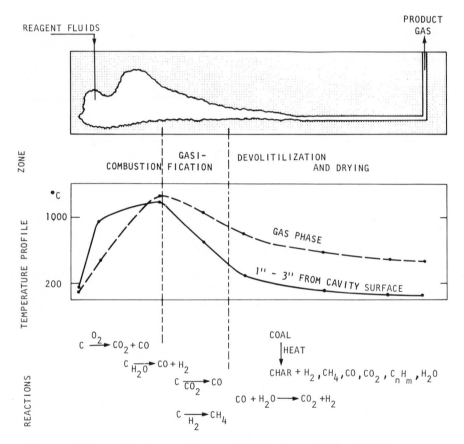

Figure 3.   *Underground coal gasification scheme (forward burn)*

perature - with the overall burning rate controlled by heat transfer. Generally, heat is generated by combustion, while heat sinks
are provided by endothermic chemical reactions, devolatilization,
evaporation of water and heat losses through conduction, convec-
tion and radiation.

To date, four gasification runs, using two blocks, have been
conducted in the simulator. The first run, primarily intended to
test system components and operating procedures, lasted for about
22 hours (during which air was injected at 13-25 m³/hr (8-15 scf/
min), and was terminated when oxygen concentrations in the product
gas exceeded 5 percent and the temperature began to fall. Table I
shows the composition of the original coal, and Table II summar-
izes the test data, which were assembled from the composition of
the product gas on the assumption that pyrolysis products made no
significant contribution to it. The contrary assumption, i.e.
that all gas was produced by thermal decomposition of the coal
(heated to 300°C at 3°C/min) would indicate a yield of only 13
scf dry gas/lb coal (0.81 m³/kg) and an energy recovery of ∿12
percent.

TABLE II

FIRST BURN AVERAGE DATA

Injection Rate	Gas Pro- duction	Heat Value	Dry Coal Effected	Material Recovery	Energy Recovery
		BTU/		Scf Dry Gas	Btu Dry Gas
SCFM	SCFM	SCF	LBS/HR	Lb. Dry Coal	Btu Dry Coal
10	12	100	15	70	65±10%

For the purpose of estimating the energy recovery and sweep
efficiency, the coal block was cut open after completion of the
test, and the burning pattern inspected. As shown in Figure 4,
the initial cross-section of the reaction channel was found to
have been enlarged to an elliptical shape (with the major axis
perpendicular to the channel), and the cavity narrowed steadily
toward the production hole. Reversal of the gas flow would
probably have allowed continuation of gasification by making for
greater sweep efficiency and thus compensating the unsymmetrical
burn cavity.

In the light of these findings, another block was prepared
for more extensive temperature measurements, and divided length-
wise into two compartments.

The second gasification test was carried out on the left half
of the test block, with the horizontal channel choked off near the
production hole in order to simulate blockage which may occur
during in-situ processing. Attempts to 'push' the burn through
the channel proved unsuccessful and, instead, as suspected from

TABLE I. COAL BLOCK ANALYSES

BASIS		Capacity Moisture	Dry	Dry Ash-Free
PROXIMATE COMPOSITION				
Moisture	%	23.6	−	−
Ash	%	4.8	6.3	−
Volatile Matter	%	29.9	39.2	41.8
Fixed Carbon	%	41.7	54.5	58.2
ULTIMATE COMPOSITION				
Carbon	%	53.6	70.2	74.9
Hydrogen	%	3.5	4.6	4.9
Sulfur	%	0.3	0.4	0.4
Ash	%	4.8	6.3	−
Moisture	%	23.6	−	−
Calorific Value, gross BTU per lb		9,200	12,050	12,860

Figure 4.   First burn cavity formation

the temperature profiles and later confirmed when the block was opened for inspection, an 'easier' path developed to the left. This eventually created a cavity in the shape and size of a large football, from which a horizontal crack extended over half the length of the block. Also, very prominent was a vertical crack above the original channel. Due to pressure limitations, air flow rates were limited to 6 scf/min injected at 5 psig. Typical product gas compositions are given in Table III. The maximum temperature was 1060°C, and product gas exited at about 100°C.

TABLE III

BURN No. 2 TYPICAL PRODUCT GAS

CO_2	CO	H_2	CH_4	C_nH_m	O_2+Ar	N_2	Heat Value BTU/SCF
7.3	6.4	10.5	2.4	0.5	13.1	59.8	88.4
12.2	6.3	9.4	2.2	0.4	7.3	62.2	80.1
11.7	7.4	11.8	2.9	0.5	7.8	57.9	100.4

After taking suitable coal samples for analysis, removing the blockage and filling the cavity with refractory cement, the third burn test was carried out on the same channel. Figure 5 shows time-data plots for this burn. Because much of the moisture had been driven forward by the preceding burn, only a relatively low-BTU product gas was produced; but by cyclical injection of steam and air, heat values could be periodically increased.

The experiment was terminated when the temperature in the adjacent coal block began to exceed 200°C.

Kinetic data for the third experiment have not yet been analyzed, but it appears that product gas heat values depend mostly on hydrogen contents, and that the use of high steam/air ratios promotes hydrogen formation via $C + H_2O \rightarrow CO + H_2$, followed by $CO + H_2O \rightarrow CO_2 + H_2$. Small increases in CH_4 content are also indicated under such conditions.

As can be observed from the temperature plots shown in Figure 6, rapid cooling of the system requires further injection of air or oxygen in order to maintain reasonably high gas heat values and minimize fluctuation in heat value.

The temperature plots presented here are only a small portion of the data collected for heat distribution calculations. Generally, the heating rate is slow up to 100°C (due to water vapourization) and thereafter becomes faster, on average approaching 0.5°C/min, and the combustion temperature usually exceeds 1200°C.

Thermocouple locations at which temperatures were measured are identified in Figure 7. The fourth burn was carried out on the right half of the block, with air as well as some CO_2 being injected for about 13 hours. The diameter of the reaction

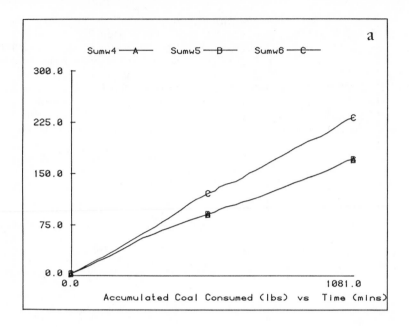

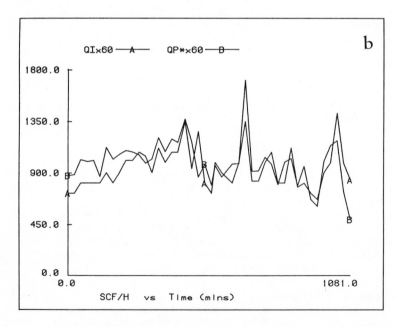

Figure 5.   Computer plots of experimental parameter, burn No. 3. (a) Coal consumption and (b) input and output gas flow rates.

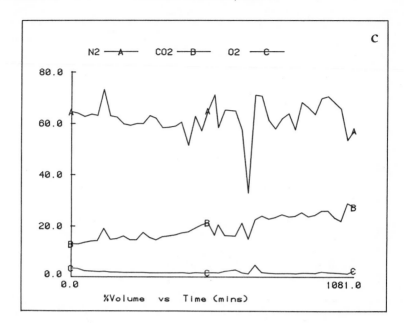

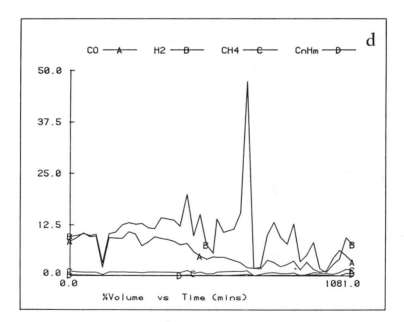

Figure 5. Computer plots of experimental parameter, burn No. 3. (c and d) Product gas composition.

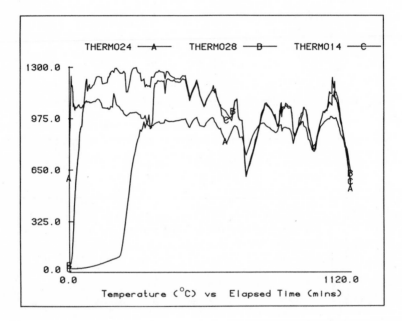

Figure 6.   Computer plots of temperature data. (Top) *Channel A, longitudinal temperature distribution: (a) ignition point; (b) 1' from ignition point; (c) 2' from ignition point.* (Top right) *Section G, radial temperature distribution.* (Bottom right) *Fluid flow: (a) inlet fluids; (b) product gas;(c)product gas off heat exchanger.*

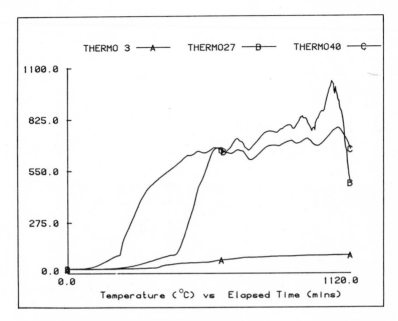

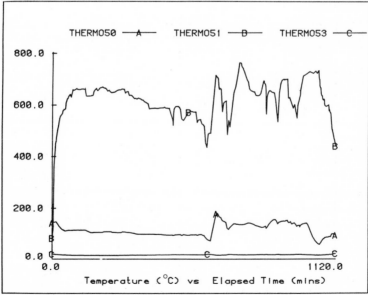

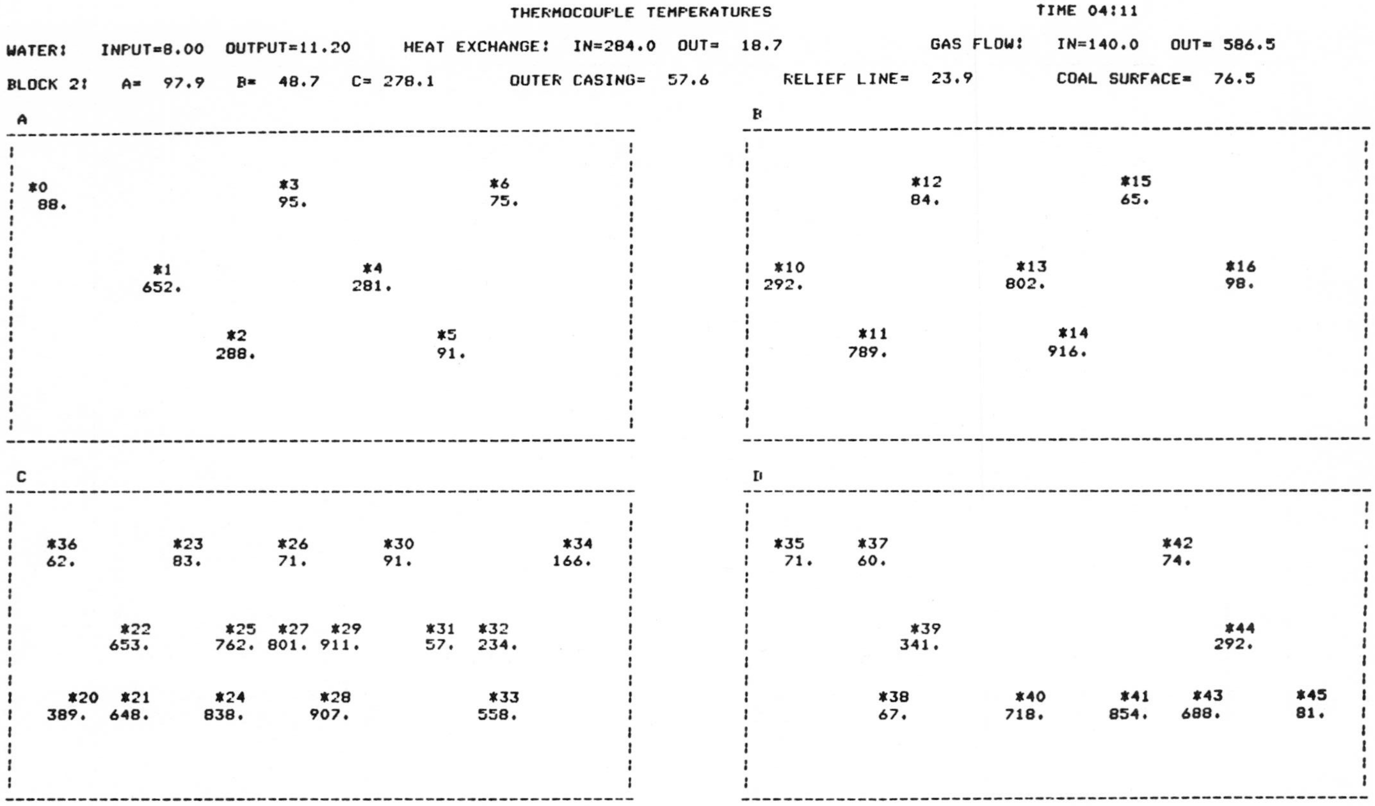

Figure 7. Location of thermocouples and temperature distribution during Burn No. 3 (longitudinal sections)

TABLE IV

PHENOLS IN TAR CO-PRODUCED BY GASIFICATION, BURN NO. 4

Sample Number

	1 (11pm)	2 (1:55am)	3 (4:30am)	Condensate
Ammonia, g/litre	2.8	3.8	3.2	6.9
Phenol, g/litre (absorption data)	4.28	6.95	6.51	-
Phenols, g/litre (identified by G.C.)	3.00	5.10	4.40	13.30
Ether-soluble organics g/litre	4.13	7.53	6.42	126.88

Weight percent distribution of identified phenols

Phenol	52.0	50.0	47.7	32.9
O-Cresol	9.4	10.0	10.2	8.4
m-Cresol	15.8	16.5	17.3	14.7
p-Cresol	12.9	12.2	12.4	11.8
O-Ethylphenol	1.3	1.7	1.7	1.8
m-Ethylphenol	1.1	1.4	1.5	3.0
p-Ethylphenol	1.0	1.3	1.4	5.0
2,6-dimethylphenol	0.2	0.3	0.5	1.4
2,4- and 2,5- dimethylphenol	2.9	3.1	3.4	9.2
2,3- and 3,5- dimethylphenol	2.4	2.4	2.6	7.7
3,4-dimethylphenol	0.9	1.1	1.1	4.2
Trimethylphenols	-	?	?	

channel was 2'' for the first one-foot length, and 7/8'' over the remaining four feet. The burn was terminated when the fire broke through the front reaction wall. Steam extinguished this fire quite effectively and produced a medium-BTU gas (with 335 BTU/scf).

It was also found that the cavity shifted toward the left wall, probably due to the fact that previous heating on that side somewhat increased the permeability of the coal.

The overall cavity geometry and channel length were similar to those found in the third test, and burning rates were also similar. Over most of the burn period (11 hrs) stable conditions prevailed, and a gas with over 110 BTU/scf was produced (CO_2-11.3%, CO-15.4%, H_2-13.3%, CH_4-1.6%, C_nH_m-0.3%, O_2+Ar-1.2%, N_2-56.9%). The air:product gas ratio was $\sim$0.73, and energy recovery reached about 70 percent (or just over 80 percent if the sensible heat of the gas is included). The heat values recorded by the calorimeter are about 8 percent higher than those computed from gas compositions and showing peaks not detected by the gas chromatograph.

Table IV shows the analysis of the (tar) condensate which was recovered from the raw product gas at a rate of 2 ℓ/hr.

Conclusion

The in-situ coal gasification simulator appears to behave much like an underground coal seam undergoing gasification, and yields typical gas at rates and efficiencies very similar to those observed in field tests. At times, it even posed similar control problems, such as locating the combustion face, gas leakage and self re-ignition.

The extensive temperature data which still remain to be correlated with char residues, the shape of the burn cavity, and chemical analyses, when combined with data from separate carbonization tests now in progress, are expected to provide information from which the extent of gasification and the sweep efficiency can be predicted with reasonable accuracy.

Much experience has also been gained which should contribute toward better understanding of the parameters that control effective gasification.

Based on this experience, an improved reactor, capable of operating at higher pressures, and allowing sampling along the reaction channel as well as flow reversal, is now being designed, and tests simulating operation of a longwall generator are planned. Future work will also include tests of hole-linking techniques and studies of the relation between cavity formation and the orientation of cleat in the coal.

Acknowledgements

I am indebted to D. Swenson and H. Wasylyk for technical assistance in the experimental work, to Dr. R.A.S. Brown for valuable advice relating to the design and operation of the

simulator facility, and to Dr. N. Berkowitz for helpful discussions and assistance in the preparation of this paper.

Abstract

In actual in-situ coal gasification, numerous processes, i.e.
oxidation, reduction, thermal cracking and a variety of catalytic
as well as non-catalytic reactions, occur in overlapping zones,
and to explore the chemistry of these reactions as single or con-
secutive unit processes is virtually impossible. It is, however,
feasible to study the individual reactions under controlled conditions by simulating in-situ gasification in the laboratory.
This paper describes a simulator which has been developed at the
Alberta Research Council and permits gasification in a two-ton
coal block. Initial gasification experiments with air, steam and
carbon dioxide are summarized, and data for product gas composition, heat propagation through the coal block, and gasification
rates as functions of the geometry of the reaction channel are
presented.

Literature Cited

1. Berkowitz, N. and Brown, R.A.S., "In-Situ Coal Gasification:
    The Forestburg (Alberta) Field Test", The Canadian Mining
    and Metallurgical Bulletin (December 1977).
2. Fitzgerald, D., 8th Arthur Duckham Fellowship Report,
    Inst. Gas. Engrs. (London) (1957), Publ. No. 516.
3. Fitzgerald, D. and van Krevelen, D.W., Fuel (1959), 38, 17.
4. Essenhigh, R.H., "Dominant Mechanisms in the Combustion of
    Coal", ASME Publ. (1971).

RECEIVED September 25, 1978.

8

Mesophase Development during Thermal Hydrogenation of an Oxidized Bituminous Coal

K. BELINKO, M. TERNAN, and B. N. NANDI

Energy Research Laboratories, Canada Centre for Mineral and Energy Technology, Department of Energy, Mines and Resources, 555 Booth Street, Ottawa, Canada

The formation of coke during carbonization of coking coals is attributed to the development of an anisotropic mesophase within the coking medium (1). The mesophase results from the alignment of certain specific pyrolysis products of the coals giving rise to liquid crystal structures. These liquid crystals separate from the carbonizing medium in the form of spheres which grow in size and ultimately coalesce to give rise to a reinforcing network within the coke. This reinforcing network is fundamental to the strength of the coke produced.

Coking coals can lose part or all of their coking character through oxidation, the extent depending on the severity of the oxidation. The incorporation of oxygen into the coal structure is thought to be primarily in the form of carbonyls and carboxylic groups (2). The presence of these functional groups in coal triggers various forms of polymerization reactions during carbonization, including the formation of oxygen cross-linkages between pyrolysis products of the coal. These polymerization reactions significantly alter the thermal rheological properties of the coal, and consequently interfere with the natural progression of carbonization. The growth of liquid crystals can be restricted by the presence of cross-linkages between the structural units of the coal and this inhibits the development of the anisotropic mesophase.

In the present work, a mild thermal hydrogenation treatment was used to revive the coking properties of an oxidized bituminous coal. The coal studied was a severely weathered high volatile bituminous coal from eastern Canada.

Experimental Procedure

An oxidized bituminous coal from Phalen Seam in Nova Scotia was obtained from the Cape Breton Development Corporation. Some properties of this coal are given in Table I. Hydrogenation was carried out in a stainless steel, vertical fixed bed reactor of 155 ml capacity (3). Approximately 150 g of sample, -4 to 8 US

TABLE I

SOME PROPERTIES OF OXIDIZED
PHALEN SEAM COAL

	wt %
Moisture	2.44
Ash	7.62
Volatile Matter	32.17
Fixed Carbon	57.77

standard sieve, were introduced into the reactor and electrolytic
hydrogen (99.9% purity) was passed up through the bed at 2.8 1/min
at S.T.P. The pressure within the reactor was held constant
during the entire run. The gas flow was initiated at the same
time as the heating cycle. The temperature of the reactor was
raised to the desired value over a period of about one hour and
held constant at that value for the duration of the experiment.
At the end of the experiment, the reaction bed was allowed to
cool in a hydrogen atmosphere. In some experiments, nitrogen was
used instead of hydrogen to confirm that the effects observed
were in fact associated with hydrogenation of the coal rather than
with high pressure effects.

The effect of reactor temperature and reactor pressure was
investigated during hydrogenation of the oxidized coal. Temperatures and pressures ranged from $375^{\circ}C$ to $475^{\circ}C$, and 0.79 MPa to
13.9 MPa respectively, with residence times of up to 3 h.

The material removed from the reactor subsequent to hydrogenation generally showed some degree of agglomeration along with
the presence of a pitch-like material. A liquid and solid yield
was recorded for each experiment. Thermal rheological properties,
such as Free Swelling Index (FSI) and dilatation, were determined
for the hydrogenated products. Oxygen content of the samples was
determined using a Perkin-Elmer (Model 240) Elemental Analyser.
Dilatation tests were performed in a Ruhr Dilatometer at a heating
rate of $3^{\circ}C/min$ to $550^{\circ}C$, according to the German Specification
DIN 51739.

Microscopic examinations of the various samples were performed on a Leitz reflected light microscope. The samples were
embedded in Lucite plastic and polished according to ASTM specifications. Micrographs were taken at a magnification of 600 using
partial crossed nicols.

Results and Discussion

Carbonization of the oxidized Phalen Seam coal at $550^{\circ}C$
resulted in a non-agglomerated char. Dilatation and FSI results
for this coal are given in Table II and confirm its non-coking
character. The semi-coke obtained from dilatation experiments
had an isotropic structure, as shown in Figure 1.

TABLE II

DILATATION AND FREE SWELLING INDEX RESULTS
FOR ORIGINAL AND HYDROGENATED PHALEN SEAM COAL

		Original Coal	Hydrogenated Coal (450oC, 13.9 MPa, 3 h)
Softening temperature	°C	339	<290
Maximum contraction	%	20	16
Maximum dilatation	%	Nil	50
Free Swelling Index		3	5½

Figure 2 is an optical micrograph of the solid residue obtained subsequent to hydrogenation of oxidized Phalen Seam coal for 3 h at 450°C and a hydrogen pressure of 13.9 MPa. Mesophase formation was indicated in the hydrogenated product. FSI and dilatation results of the hydrogenated coal were found to improve considerably over those of the original coal and the semi-coke was found to be agglomerated and hard. Figure 3 is an optical micrograph of the semi-coke obtained from dilatation experiments with the hydrogenated oxidized coal, showing a flow-type coke structure.

The effect of reactor temperature and hydrogen pressure on the weight yield of solid residue recovered subsequent to hydrogenation of the oxidized Phalen Seam coal is shown in Figure 4. The yield of solid residue decreased as the hydrogen pressure in the reactor increased, indicating a higher conversion of the coal into liquid and gaseous products. A similar trend was observed for increasing reactor temperature.

Oxidized Phalen Seam coal hydrogenated at 450°C for 3 hours at different hydrogen pressures gave the dilatation results presented in Table III. Optical micrographs of the semi-cokes obtained from dilatation experiments are shown in Figures 5 to 8.

TABLE III

DILATATION RESULTS FOR OXIDIZED PHALEN SEAM
HYDROGENATED FOR 3 HOURS AT 450°C

		Pressure, MPa				
		1.5	3.5	7.0	10.4	13.9
Softening temperature	°C	NS*	NS*	412	350	<290
Maximum contraction	%	Nil	Nil	3	19	16
Maximum dilatation	%	Nil	Nil	Nil	5	50
Plasticity index		Nil	Nil	0.06	0.29	0.22

* NS - No softening

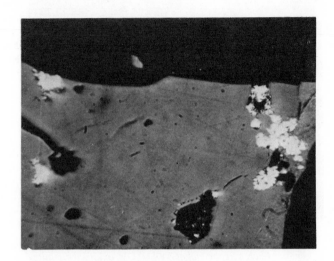

Figure 1. Optical micrograph of semicoke from carbonization of oxidized Phalen Seam coal at 550°C, showing isotropic coke structure

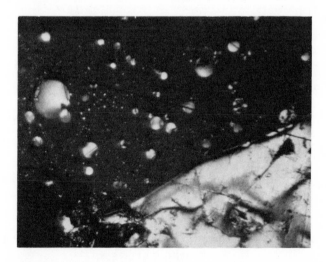

Figure 2. Optical micrograph of solid residue recovered from hydrogenation of oxidied Phalen Seam coal, showing liquid crystal spheres; 450°C, 13.9 MPa, and 3.0 hr

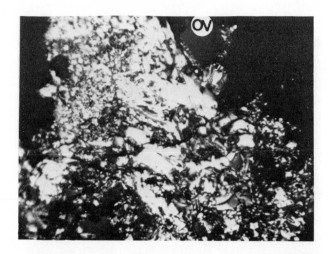

Figure 3.   Optical micrograph of the semicoke from carbonization of the hydrogenated Phalen Seam coal at 550°C; OV, oxidized vitrinite

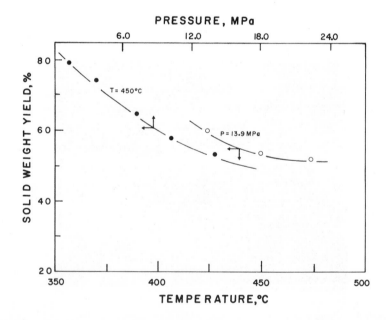

Figure 4.   Weight yield of solid residue as a function of reactor temperature and
pressure for hydrogenation experiments carried out for 3 hr with oxidized Phalen
Seam coal

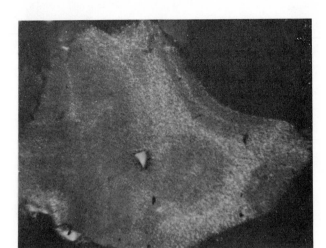

Figure 5.   Semicoke of oxidized Phalen Seam coal hydrogenated at 450°C and 1.5 MPa for 3 hr, showing a very fine grain, mosaic coke structure

Figure 6.   Semicoke of oxidized Phalen Seam coal hydrogenated at 450°C and 7.0 MPa for 3 hr, showing fine grain, mosaic coke structure

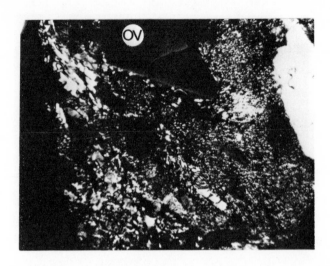

Figure 7.   Semicoke of oxidized Phalen Seam coal hydrogenated at 450°C and
10.9 MPa for 3 hr, showing coarse grain, mosaic coke structure

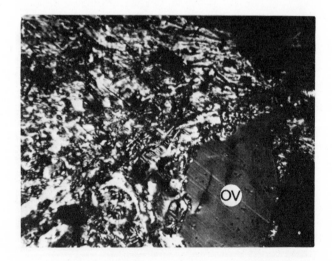

Figure 8.   Semicoke of oxidized Phalen Seam coal hydrogenated at 450°C and
13.9 MPa for 3 hr, showing flow-type coke structure

A very fine grain mosaic coke structure was observed for coal samples hydrogenated at pressures of 1.5 MPa and lower (Figure 5). The structure of the semi-coke varied from a fine grain mosaic for coal samples hydrogenated at 7.0 MPa, to a coarse grain mosaic for coal samples hydrogenated at 10.9 MPa, Figure 6 and 7 respectively. A predominantly flow-type coke structure was observed in the semi-coke of coal samples hydrogenated at 13.9 MPa, as shown in Figure 8.

The oxygen content of the solid residue recovered from hydrogenation experiments of Phalen Seam coal at various hydrogen pressures are given in Table IV. The oxygen content was found to decrease from 6.63% for coal samples hydrogenated at 1.5 MPa to 2.54% for coal samples hydrogenated at 13.9 MPa.

TABLE IV

OXYGEN CONTENT OF PHALEN SEAM COAL
HYDROGENATED AT 450°C FOR 3 HOURS
AT DIFFERENT HYDROGEN PRESSURES

	Oxygen, wt %
1.4 MPa	6.63
3.5 MPa	6.22
6.9 MPa	3.74
13.9 MPa	2.54

The conversion of a non-coking coal to a coking coal under hydrogenation conditions has been reported previously in the literature with little reference made to the mechanism of the process (4,5,6,7,8). From the results reported here, it appears that hydrogenation removes some of the limitation on the growth of liquid crystal spheres, thereby enabling the development of an anisotropic mesophase during carbonization.

Hydrogenation of oxidized Phalen Seam coal resulted in deoxygenation of the coal, as shown in Table IV. Infra-red analysis has revealed that the hydrogenated product has a significantly lower proportion of aromatic OH- and C-O groups, and C-O-C groups than the oxidized coal, and a higher proportion of $-CH_2$ and $-CH_3$ groups (4). It is possible that removal of oxygen from the coal during hydrogenation improved the thermal rheological properties of the coal. It has previously been reported that the presence of oxygen and sulphur in a carbonizing medium can lead to the formation of cross-linkages between pyrolysis products (9,10). These cross-linkages restrict the growth of liquid crystals, thereby destroying the coking properties of the carbonizing medium. Oxygen removal from an oxidized coal would therefore not only improve the thermal rheological properties of the coal, but would lessen some of the inhibition on the growth of liquid crystals which can be brought about by the formation of oxygen cross-linkages.

Conclusions

Thermal hydrogenation of an oxidized bituminous coal under appropriate conditions can restore the coking properties of the coal. Removal of oxygen from the oxidized coal during hydrogenation improved the thermal rheological properties of the coal, and decreased the likelihood of oxygen cross-linkage formation between pyrolysis products of the coal. Liquid crystals are therefore allowed to form with a lesser degree of limitation on their growth thus enabling the development of an anisotropic mesophase during carbonization.

Acknowledgement

The authors gratefully acknowledge the technical contributions of S.E. Nixon, B.H. Moffatt and I. Johnson during the course of this work.

Abstract

A severely weathered bituminous coal from eastern Canada was treated by thermal hydrogenation under various reactor conditions. The coking properties of this coal were found to be restored under appropriate hydrogenation conditions. The semi-coke of the hydrogenated coal exhibited an anisotropic coke structure. The size of the anisotropic domains in the semi-coke was found to depend on reactor temperature and hydrogen pressure during hydrogenation.

Literature Cited

1. Marsh, H., Fuel (1973) 52, 205.
2. Mazumdar, B.K., Chakrabarty, S.K., Saha, M., Anand, K.S. and Lahiri, A., Fuel (1959) 38, 469.
3. Ternan, M., Nandi, B.N. and Parsons, B.I., Mines Branch Research Report R-276, Department of Energy, Mines and Resources, Ottawa; (1974).
4. Nandi, B.N., Ternan, M., Parsons, B.I. and Montgomery, D.S., 12th Biennial Conference on Carbon, American Carbon Society, (1975) 12, 229-232.
5. Lander, C., Sinnatt, F.S., King, J.G. and Crawford, A., British Patent Specifications 301-720; (1928).
6. Crawford, A., Williams, F.A., King, J.G. and Sinnatt, F.S., Fuel Research Tech. Paper 29, Dept. Sci. and Indust. Research, His Majesty's Stationary Office, London; (1931).
7. Samec, M., Proceedings of the Symposium on the Nature of Coal, Central Fuel Research Institute, Jealgora, India; (1959).
8. Ahuya, L.D., Kini, K.A., Basak, N.G. and Lahiri, A., Proceedings of the Symposium on the Nature of Coal, Central Fuel Research Institute, Jealgora, India; (1959).

9. Kippling, J.J., Shooter, P.V. and Young, R.N., Carbon (1966) 4, 333.
10. Nandi, B.N., Belinko, K., Ciavaglia, L.A. and Pruden, B.B.; Fuel (1978) 57, 265.

RECEIVED September 25, 1978.

Perpetual and Renewable Sources

Methane Production from Manure

H. M. LAPP

Department of Agricultural Engineering, University of Manitoba,
Winnipeg, Manitoba, Canada

Recent general public awareness that the reserve supply of
conventional liquid and gaseous fossil fuels is rapidly diminishing has given rise to a serious concern for securing the future
energy supply for all sectors of society. This major concern was
brought into sharp focus by the oil embargo imposed in 1973 by
the Organization of Petroleum Exporting Countries (OPEC). This
event has been of such significance that energy considerations
are frequently time related as to their occurrence having taken
place "Prior to" or in "Post" OPEC years. Programs in Canada to
promote energy conservation and to stimulate research into potentially viable alternate sources for energy fuels have been very
active during the past five "Post OPEC" years. The utilization
of the anaerobic digestion process for the conversion of organic
raw material into a useful energy fuel has emerged as one technology to receive serious attention. This paper will discuss the
production of methane from manure.

Anaerobic digestion systems have been installed in the agricultural sector of various countries, the majority of which lie
within the tropics. Po (1) reported that 7500 units were operational in Taiwan and in a recent personal communication he confirmed that more than 8000 units were currently operational. The
technology has been promoted vigorously in the People's Republic
of China since 1970 and Smil (2) has reported that the number of
systems, ranging in capacity from a few up to 100 cubic metres
has reached 4.3 million. The Institute of Agricultural
Engineering and Utilization in Korea (3) reported that 24000
units were installed between 1969 and 1973. India has promoted
the installation of biogas plants following the initiation of experiments in 1939. Adoption of the technology has expanded with
the encouragement of the Khadi and Village Industries Commission.
The Gobar Gas Research Station was started in Ajitmal, Etawah
(Uttar Pradesh) in 1961 and a variety of designs for gas plants,
developed by Singh (4), were published by this Institution in
1971. Many of these plants have been built in rural India and it
is expected that upwards of 200,000 units will soon be operational.

A limited number of digesters were built in Europe after World
War II to use manure for the production of methane gas for farm
fuel. The use of these units were generally discontinued about
1960 as low cost liquid petroleum fuel again became available.
Findlay (5,6) following two surveys of 95 Gobar Gas Plants installed in Nepal in 1975 and 1976 reported that gobar gas technology was being well accepted in Nepal. The Nepal Agricultural
Bank, which provides low interest loans for gas plant construction, have forecast that many thousand plants will be built in
Nepal in the immediate future. Interest in the potential use of
anaerobic digesters to produce methane from manure in Canada has
increased in recent years.

A program was initiated in 1971 at the University of
Manitoba to investigate the technical and economic feasibility
of producing methane gas from animal manure employing the anaerobic digestion process. Technical and economic feasibility for
on-farm operation of anaerobic digesters in Canada is constrain-
ed by the occurrence of cold temperatures in winter. However,
modern confinement housing systems for livestock enterprises result in rapid and concentrated collection of large quantities of
manure which is favorable to the successful operation of anaerobic digesters. Animal manure contains large quantities of organic matter which will yield significant quantities of methane
gas when subjected to bioconversion through anaerobic digestion
in a controlled environment. The program at the University of
Manitoba has involved laboratory investigations since 1971 and
the operation of a pilot plant using swine manure since 1973.
Two farm scale demonstration systems are under construction in
Canada, one in Manitoba on the W. Langille Farm at Stonewall and
one in Ontario on the John Fallis farm at Peterborough.

A number of universities, research stations, farms and feed
lots in the United States have laboratory programs, pilot plants,
farm units and commercial systems in various stages of investigation, development and operation. A selected number of these
activities in the Midwestern United States are described in a
1978 travel report by Lapp and Buchanan (7).

Anaerobic Digestion

Production of combustible gas is a naturally occurring biological process involving the decomposition of organic matter.
It was discovered in the 17th Century when scientists observed
the so-called "marsh gas" burning on the surface of swamps.
Natural gas was originally formed by the decomposition of organic materials from prehistoric plants and animals that had become
trapped in sediments. The biological process has become known
as anaerobic digestion with which the production of gas is associated. This gas, consisting primarily of methane (CH_4) and
carbon dioxide (CO_2) is commonly referred as "biogas" and it
differs only from "marsh gas" and "natural gas" in its degree of

purity with respect to its methane content. The production of
biogas, often referred to directly as methane, from animal manure
requires an understanding of anaerobic digestion as well as the
associated manure and biogas handling requirements.

Anaerobic digestion means the biological breakdown of organic matter in the absence of oxygen. It is a two-stage process
in which each stage is performed by a distinct group of bacteria.
During the first stage a complex mixture of fats, carbohydrates
and proteins in manure are degraded and converted to simple organic acids such as acetic and propionic. The "acid-forming"
bacteria reproduce rapidly and are not sensitive to changes in
their environment. The second stage of the process involves the
conversion of the organic fatty acids into methane and carbon
dioxide. The "methane-forming" bacteria are relatively few in
number, do not reproduce rapidly and are extremely sensitive to
their environment and particularly to the presence of oxygen.
This two stage process is represented chemically by

$$(C_6 H_{12} O_6)_n \rightarrow 3n\ CH_3\ C\ O\ O\ H$$

$$CH_3\ C\ O\ O\ H \rightarrow CH_4 + CO_2$$

$$CO_2 + 4H_2 \rightarrow CH_4 + 2H_2O$$

$$CO_2 + H_2O \rightarrow H_2\ CO_3$$

$$(C_6 H_{12} O_6)_n + 3n\ H_2O \rightarrow 3n\ CH_4 + 3n\ H_2\ CO_3$$

The acid-forming and methane-forming bacteria work simultaneously in the anaerobic digestion process. Oxygen is excluded from the digester and various other environmental conditions
are controlled so that the methane-formers are maintained in
balance with the acid-formers. Otherwise, the methane-formers
would be inhibited and in some cases would cease to function entirely.

Technical Feasibility of Anaerobic Digestion
for Methane Production

Three major system components which need to be examined
when assessing the technical feasibility of using anaerobic
digestion to produce methane gas from animal manure include
i) manure handling ii) biological process stability and iii)
biogas handling.

Manure Handling. The physical characteristics of manure
are such that manure handling systems are difficult to engineer
with any degree of precision. Proven manure-handling systems
will need to be integrated with anaerobic digestion systems if
they are to be accepted by a livestock enterprise operator.

Energy recovery schemes which are not reliable or labor-efficient in all aspects of manure handling will be abandoned quickly be a livestock producer.

Manure is excreted from animals at near optimum temperature for anaerobic digestion and optimum biogas production. Location of a digester inside the confines of a livestock building to receive excrement without heat loss would be desired. Unfortunately the explosive characteristics of biogas has prevented such a development from taking place. Consequently, fresh manure either positively pumped, mechanically scraped or hydraulically flushed from a confinement building to an adjacent digester appears necessary for attaining technical feasibility from a manure handling standpoint.

Handling does not end with entry into the digester since contents must be mixed, scum formation prevented and effluent removed to disposal or storage. Welsh et al (8) reported that anaerobic digestion renders the effluent less offensive and more pumpable with fewer settling problems.

Biological Process Stability. The anaerobic digestion process involves a complexity of interrelated biochemical reactions, many of which are not clearly understood. Anaerobic digestion therefore, has a poor reputation from a process-stability viewpoint. Municipal digesters are frequently attended by skilled operators with elaborate monitoring facilities at their disposal. The adoption of on-farm anaerobic digestion systems for methane production are unlikely to occur under such requirements for successful operation.

Environmental factors effecting process-instability include sudden changes in temperature, loading rates, the nature of the organic material and the presence of toxic elements. A further complexity to the stability of anaerobic digesters handling animal manure is due to the fact that these digesters operate outside the range of chemical parameters considered "normal" for municipal-sludge digestion. Parameters commonly used to monitor process stability include pH, alkalinity and the concentration of volatile acids. Municipal sewage treatment digesters typically operate at volatile acids concentrations lower than 1000 mg L^{-1} as acetic acid with an alkalinity range of 1000 to 5000 mg L^{-1} as Ca CO_3 and a pH range of 6.6 to 7.6. Excessive concentrations of volatile acids and ammonia nitrogen have been considered toxic to methane formers.

A range of chemical parameters which have been recorded during steady-state digestion of swine manure during pilot plant studies at the University of Manitoba are recorded in Table 1. It is significant to note that process stability was maintained even though the chemical parameters recorded were well outside the normal range for municipal digesters. This occurrence demonstrates the unique nature of animal-manure digestion as compared to municipal sewage.

Table I. Digester Operating Characteristics

Chemical Parameter	Digester			
	A	B	C	D
Total volatile acids (Mg/l as HAc)	850–4680	1050–4350	1050–5400	1050–4750
Alkalinity (mg/l as $CaCO_3$)	7480–16600	8000–16850	8630–17230	8250–16330
pH	7.80–8.30	8.0–8.20	7.70–8.30	7.70–8.30
Total ammonia (mg/l as N)	2260–3580	2240–3530	2330–3570	2430–3620

Lapp et al (9) and Kroeker et al (10) during pilot plant and laboratory studies at the University of Manitoba have demonstrated that extreme process stability is possible in a digester using swine manure. Digester-instability reported for swine and poultry manure by Anthonisen and Cassell (11); Hart (12); Gramms et al (13) and Schmid and Lipper (14) need not occur if proper attention is given to acclimating the methane-forming bacteria to the manure slurry substrate. This stability is due to the relatively high concentrations of ammonia nitrogen in solution and is possible only after acclimation of methanogenic bacteria to the high nitrogen content of the slurry. Stable digester operation was maintained throughout the experiments in spite of large temperature fluctuations at high and low organic loading rates, and with large daily variations in organic loading rates. All of these conditions would be typical of a full-scale installation on a farm.

1. Seeding. This is a component of the start-up practice and consists of the addition of actively digesting material to a newly operating digester. The addition ensures that a culture of methane-producing bacteria is present for start-up.

2. Nutrient Balance. All biological systems require an adequate supply of nutrients, particularly nitrogen, phosphorous, and potassium. Animal manure normally contains an adequate, well-balanced nutrient supply to support the existence of a thriving biological system. For good anaerobic digestion the carbon to nitrogen ratio should range between 15 to 1 and 30 to 1.

3. Volatile Solids. Volatile solids represents the organic component of the total solids present and approximately 85 percent of the total solids are volatile. A biological system always converts a portion of its substrate into new cell mass and generally less than 50 percent of the volatile solids are

destroyed in practice.

4. Start-up. Methane-forming bacteria are present in most manure handling systems but it takes them a long time to multiply
into an efficient methane-producing population. The time to
establish a satisfactory population of bacteria can be reduced
by adding an actively digesting material (seed) from another digester. At least 15% of the volume of the digester should be
filled with seed at start-up. The seed should be added to the
digester which has already had the remaining volume filled with
water warmed to the intended operating temperature. Fresh manure can then be added but slowly at first (approximately 10% of
the planned daily load) to allow the bacteria time to acclimate
to the new environment of the fresh manure. After gas production
has reached about 50% of that expected for the low loading rate,
the loading rate should be increased gradually over a three week
period before reaching the desired loading rate. This procedure
should insure good gas production in about four weeks from start-
up.

5. Loading Rate. Loading rate is expressed in terms of the mass
of volatile solids added per unit of digester volume. Accepted
loading rates range from 0.7 to 5.0 kg per cubic metre per day
(0.6 to 0.31 lb/ft^3/day).

6. Retention Time. The solids retention time represents the
average time that microorganisms remain in the system and it can
be determined by dividing the mass of volatile solids in the digester by the mass leaving the system per day. The retention
time must be great enough to allow time for the methane-formers
to convert the acids to biogas. Normally 15 to 30 days are adequate for manure digestion.

7. Temperature. Two temperature ranges exist for good biogas
production, mesophilic and thermophilic. Most digesters are operated within the mesophilic range of 20° to 45°C (69° to 113°F).
The thermophilic range is 45 to 55°C (113 to 131°F).

8. Alkalinity and pH. Pilot-plant digesters at the University
of Manitoba have operated successfully at pH levels up to 8.5
and at alkalinities ranging up to 14000 mg/L. These levels are
well above those normally found in municipal digesters in which
pH ranges of 7.2 to 7.6 and alkalinities of 1000 to 5000 mg/L
normally occur.

9. Mixing. Mixing can be accomplished by mechanical recircula-
tion, agitation or by controlled gas flow methods. The practice
is desired to facilitate on intimate contact between methane
forming bacteria and their substrate and to prevent the formation of surface scum in the digester.

10. Total Solids. The optimum total solids present in an anaerobic digester should normally range from 7 to 9 percent. Animal
manure contains from 10 to 25 percent dry solids and may require
dilution prior to loading into an anaerobic digester.

Biogas Handling. Quantitative biogas production potential

Figure 1. *Pilot plant for biogas production at the Faculty of Agriculture's Glen-lea Research Station, University of Manitoba, Winnipeg, Canada*

Figure 2.   A 16,000 poultry layer operation in a converted dairy barn with an anaerobic-digestion biogas plant in the attached shed in the foreground (W. Gibbons' farm at Ripon, Wisconsin)

Figure 3.   Biogas-fired boiler installed at the Calorific Recovery Anaerobic Process Plant, Guymon, Oklahoma and operated by Thermonetics Incorporated, Oklahoma City.   Manure is processed from feed lots housing 100,000 head of beef cattle.

from manure is related to the biodegradable organic matter (volatile solids) present in the manure. Volatile solids (VS) are expressed in units of kg of volatile solids per day per 1000 kg of liveweight. However the volatile solids from one species, say poultry, is often more biodegradable than from another species, say beef cattle. Biodegradability of manure is also affected by the length and type of storage which occurs prior to digestion. Typical values used to estimate potential biogas production from livestock manure are contained in Table II.

Table II.   Typical Values for Estimating Potential Biogas Production From Various Types of Livestock

	Growing-Finishing Swine	Dairy Cows	Laying Hens	Beef Feeders
Undiluted Fresh Feces plus Urine L day^{-1} 1000 kg^{-1} Liveweight	65	82	53	60
Volatile Solids Production Rate kg VS day^{-1} 1000 kg^{-1} liveweight	4.8	8.6	9.5	5.9
Fraction of Volatile Solids Converted to Biogas	0.50	0.35	0.60	0.45
Biogas Production m^3 day^{-1} 1000 kg^{-1} liveweight	2.62	3.28	6.21	2.66
m^3 day^{-1} m^{-3} of digester	1.1	1.1	1.3	1.3

Methane is flammable and when mixed with air in proportions ranging from 5 to 15 percent by volume, is explosive. Safety regulations covering buildings, electrical and mechanical equipment installations should be strictly adhered to during the planning, construction and operation of an anaerobic digestion system.

Biogas is normally composed of 60 to 70 percent methane and from 30 to 40 percent carbon dioxide with small amounts of hydrogen sulfide and other impurities. Because its major constituent is methane its properties closely approximate those of pure methane. The critical pressure of methane is 4710 kPa at −82.3°C. It is referred to as a permanent gas since it cannot be liquified by pressure at ordinary temperatures. This property gives rise to a storage problem associated with production and utilization of biogas in regions like Canada with extreme climates. In such locations the summer season is most favorable to biogas production while a major need for space heating occurs in winter. Biogas storage from summer to winter seasons is not practical. An ideal farm system would involve a production capacity matched to

an energy utilization requirement. Should the utilization requirement involve the operation of internal combustion engines,
which is technically feasible, additional gas cleaning technology to remove hydrogen sulfide is required and if efficiency is
to be increased then carbon dioxide should also be removed.

Summary

     The production of methane from animal manure is technically
feasible under the management of a livestock enterprise operator.
Many small farmers are successfully operating small scale digesters in tropical countries.
     The decision to build an anaerobic digester on a Canadian
livestock farm should be based on advantages to be derived from
the digester as a component of the total manure handling system.
Economic justification can only be supported at present if credits are given to materials handling, environmental improvement,
fertilizer nutrient retention, pollution reduction and biogas
(methane) production. Economic feasibility for large scale confinement housing systems to employ anaerobic digestion for biogas production is near at hand in the cold climate areas of
North America. This viability is being accelerated as prices
for conventional liquid petroleum fuels continue to escalate and
as urban environmental concerns continue to grow.

LITERATURE CITED

1. Po, C. Proc. of Int. Biomass Energy Conf. (1973). Pub. by
    the Biomass Energy Institute Inc., Winnipeg, Canada. Pro-
    duction and Use of Methane From Animal Wastes in Taiwan.
    Pp. XVI - 1 to XVI - 8.

2. Smil, V. Environment (1977). 19 (7). Energy Solution in
    China. Pp. 27:31.

3. Institute of Agricultural Engineering and Utilization (1973).
    Present Status of Methane Gas Utilization as a Rural Fuel
    in Korea.

4. Singh, R. B. Gobar Gas Research Station, Ajitmal, Etawah
    (U.P.), India (1971). Biogas Plant, Generating Methane
    from Organic Wastes.

5. Findlay, J. H. Development and Consulting Services, Butwal,
    Nepal (1976). Report on First Inspection Visit to 95
    Gobar Gas Plants Constructed in Nepal.

6. Findlay, J. H. Development and Consulting Services, Butwal,
    Nepal (1977). Report on Second Inspection Visit to 95
    Gobar Gas Plants Constructed in Nepal.

7. Lapp, H. M. and L. C. Buchanan. <u>Agriculture Canada</u>, <u>Ottawa</u>
    (1978). A Travel Report on a Study of Methane Production
    From Animal Manure in the Midwestern United States. 44 p.

8. Welsh, F. W., D. D. Schulte, E. J. Kroeker and H. M. Lapp.
    <u>Can. Agric. Eng.</u> (1977). 19 (2). The Effect of
    Anaerobic Digestion Upon Swine Manure Odours. Pp. 122 –
    126.

9. Lapp, H. M., D. D. Schulte, E. J. Kroeker, A. B. Sparling
    and B. H. Topnik. "Managing Livestock Wastes". <u>Amer.</u>
    <u>Soc. Agric. Engrs</u>. St. Joseph, Mich. (1975). Start-up
    of Pilot Scale Swine Manure Digesters for Methane
    Production. Pp. 234 – 238, 243.

10. Kroeker, E. J., H. M. Lapp, D. D. Schulte, J. D. Haliburton
    and A. B. Sparling. <u>Can. Soc. Agric. Eng.</u>, Annual Meet-
    ing, Halifax. 1976. Unpublished Paper No. 76-208,
    Methane Production From Animal Wastes II – Process
    Stability.

11. Anthonisen, A. and E. A. Cassell. <u>New York State</u>
    <u>Department of Health</u>, Albany, N.Y. (1966). Studies on
    Chicken Manure Disposal. II Anaerobic Digestion, Research
    Report No. 12. Pp. 65 – 112.

12. Hart, S. A. <u>Jour. Water Poll. Control Fed.</u> (1963). <u>35</u>.
    Digestion Tests of Livestock Wastes. Pp. 748 – 757.

13. Gramms, L. C., L. B. Polkowski and S. A. Witzel. <u>Trans.</u>
    <u>Amer. Soc. Agric. Engrs</u>. (1971). <u>14</u>. Anaerobic
    Digestion of Farm Waste (dairy bull, swine and poultry).
    Pp. 7 – 11, 13.

14. Schmid, L. A. and R. I. Lipper. <u>Proc. Conf. on Agric. Waste</u>
    <u>Management</u>, Cornell Univ., Ithaca, N.Y. (1969). Swine
    Wastes, Characterization and Anaerobic Digestion. Pp. 50-
    57.

RECEIVED September 25, 1978.

10

Liquid Fuels from Carbonates by a Microbial System

MORRIS WAYMAN and MARY WHITELEY

Department of Chemical Engineering and Applied Chemistry,
University of Toronto, Toronto, Ontario M5S 1A4

In his review of techniques for the enrichment, isolation and maintenance of the photosynthetic bacteria, van Niel (1) contrasts two kinds of cycles of matter, a "primitive cycle" in which sulfide is oxidized to sulfate by photosynthetic bacteria, the resulting sulfate being reduced to sulfide again by other, non-photosynthetic microbes; and a "terrestrial cycle" in which green plant photosynthesis provides an organic base for biological regeneration of carbon dioxide. He illustrated the two cycles as shown in Figure 1. Carbon dioxide is essential in both cycles, being fixed in photosynthesis and released in fermentation. To the extent that the two cycles interchange CO_2, they are interactive. However, the two cycles differ with respect to some portions of the sun's radiation utilized by the red bacteria and the green plants, which may be significant in the functioning of the ecosystem as a whole. The photosynthetic bacteria are autotrophs, whereas the sulfate reducers live on the organic matter formed by them. Plant, animal and most microbial oxidations require molecular oxygen, whereas microbial sulfate formation from sulfide takes place anaerobically, without the use of molecular oxygen.

Photosynthetic bacteria and sulfate reducing bacteria are found in nature in close association (1,2), both groups being engaged in the sulfur cycle. Upon isolation of the pure cultures, the recycling of sulfur compounds between them is upset. Membrane-separated culture offers an opportunity to study the growth and metabolism of each of the two kinds of organisms, and at the same time the interaction between them (3). This paper reports on such interactions in membrane-separated anaerobic culture of a red photosynthetic bacterium and a colourless nonphotosynthetic sulfate reducing bacterium. In this microbial system, carbonate was the only source of carbon for growth.

The growth of the separate species has been followed, as well as the formation and disappearance of sulfide. The harvested microbes were analysed for protein content.

Materials and Methods

Organisms. Both new isolates were obtained from a red-purple
bloom which occurs in Durum Lake, Saskatchewan. Durum Lake is
rich in sodium sulfate, the heavier layers containing about 7 to 8
percent of this salt. The bloom occurs at a lower concentration,
about 0.35 percent, 1 to 3 cm below the surface. It contains many
different microorganisms, including algae and protozoa. A
Chlorella isolated from the bloom has been studied separately (4).
The separation of microorganisms into useful groups was begun
using a Winogradsky column and low incandescent illumination. The
upper layers were green and the lower layers red-purple.
    The interdependence of the photosynthetic bacteria and the
sulfate reducing bacteria was observed during the enrichment and
isolation process. As the pH of Pfennig's standard medium (5) was
raised there was an increase in the red photosynthetic bacteria in
the crude enrichment cultures. Richly coloured red cultures were
obtained at the higher pH values, and this reached a maximum at
initial pH 9.0. Using so-modified Pfennig's medium, cultures of
the red photosynthetic bacteria could be directly isolated from
the bloom samples without pre-enrichment in a Winogradsky column.
Green photosynthetic bacteria could also be isolated in this modified medium, but only from a Winogradsky column. Following growth
of the red bacteria in this medium, the crude cultures were found
to have pH 7.5 - 7.8, as the medium was not buffered to pH 9.
    In order to obtain pure cultures, the crude culture which con-
tained Chromatium species was serially diluted in this modified
medium as suggested by van Niel (1). This was repeated several
times to ensure the purity of the culture. The agar shake method
was not used as the large photosynthetic bacteria fail to grow in
agar shake tubes (1). In pure culture the Chromatium species did
not grow as well as in the mixed culture situation. The culture
was not as dense or as rich in colour as the crude cultures.
    The red photosynthetic bacterium appears to be identical mor-
phologically and in its behaviour with a stock culture of
Chromatium warmingii obtained from the American Type Culture
Collection, No. 14 959. It will be referred to as Chromatium
warmingii NI (NI = new isolate).
    The sulfate reducing bacterium was isolated from the crude
culture by pre-enrichment in a lactate-sulfate broth (1). Several
anomalous properties of this sulfate reducer were noted. It grew
well on Medium C of Butlin (6), plus agar, and formed black colonies, but failed to grow on subculture. The agar around the
colonies showed blackening. This blackening, and the black colonies, imply sulfate reduction. The culture was incubated in an
anaerobe jar with a H_2-CO_2 atmosphere. Cotton wool soaked with
lead acetate was placed in the jar to prevent H_2S poisoning of the
Pd catalyst (7). Black lead sulfide became evident.
    Structurally, the sulfate reducing bacteria are pointed, non-
motile rods, often associated in pairs. They are gram-negative
and non-sporulating. These properties do not correspond with the

descriptions of the sulfate reducing bacteria belonging to the
Desulfovibrio or Desulfomaculum groups. However, N. Hvid-Hansen
has described (8) a sulfate reducing bacterium which he isolated
from sulfide containing waters characterized by a slightly alkaline
reaction, high total solids and high bicarbonate, low sulfate and
a low content of other di- and tri-valent ions. This description
matches rather closely to the modified Pfennig medium we used, and
our sulfate reducers were able to grow in this medium alone.
Hvid-Hansen reported a failure of his organisms to grow on subculture in media suitable for the growth of Desulfovibrio. He named
his organism Desulforistella hydrocarbonoblastica, the specific
name being chosen because the organism was found to contain bituminous substances. Our sulfate reducing bacterium was found to
grow, and was maintained, on the medium of van Niel for photosynthetic microorganisms, plus agar. Whether or not our sulfate
reducer is Desulforistella hydrocarbonoblastica is questionable,
and we cannot prove it since the original culture has been lost.
However, due to the many similarities we have tentatively called
it Desulforistella sp.

Media. The media employed were based on Pfennig's standard
medium (5) for the growth of photosynthetics at pH 6.8. It was
modified for other experiments to pH 9.0. At this pH a precipitate formed which was filtered off with Whatman No. 1, followed by
passage through a Millipore filter for sterilization. All cultures
were grown at room temperature (21 ± 1°C).

Illumination. Chromatium cultures were illuminated continuously by three 24 watt incandescent bulbs at a distance of 25 to
30 cm.

Membrane-Separated Culture Apparatus. One of the advantages
of membrane-separated cultures is that the extent of cell growth
of each of two (or more) cultures can be conveniently measured,
and interaction can be observed at the same time (3). It was
therefore decided to use this type of apparatus to study the interaction of Chromatium warmingii and Desulforistella sp. A diagram
of the apparatus is shown in Figure 2. A commercial membrane-
separated culture apparatus (Belco Glass Co.) was adapted to acco-
modate for anaerobic conditions and easy sampling. The domed lids
of the original apparatus were replaced by flat aluminum lids so
that the fermenters could be completely filled. In addition, extension tubes were attached which reached a level above the lids
for filling and sampling. The apparatus was filled through these
extension tubes through Millipore filters. An overflow tube in
the top of each lid allowed the apparatus to be filled to the top.
Each fermenter - each side - held about 1700 ml. After the apparatus was filled and inoculated, sterile Millipore filters were
attached to the extension tubes. These were used to feed small
amounts of media into the unit via a syringe. This addition caused

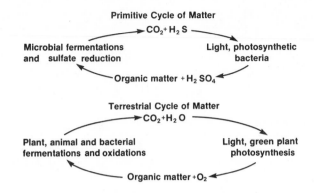

Figure 1. *Primitive cycle of matter*

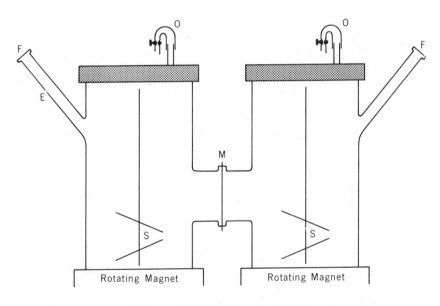

Figure 2. *Membrane-separated culture apparatus, modified for anaerobic operation. M, nucleopore membrane; O, overflow; S, magnetic stirrers; E, extension tubes; F, millipore filters.*

equal volumes of the cultures to overflow, and these were collected
from the overflow tubes. Approximately 10 ml samples were taken
for each reading. The membrane used to separate the cultures was
a Nuclepore membrane, pore size 0.2 μm.

Growth Rates. The following membrane-separated culture systems were set up, and during the course of growth, dry weight of
microorganism and the pH of the cultures were followed:

Side 1 Fermenter	Side 2 Fermenter	Initial pH
C. warmingii (ATCC)	Desulforistella sp.	6.8
" " " (ATCC)	" " "	9.0
" " " (NI)	" " "	6.8
" " " (NI)	" " "	9.0
" " " (ATCC)*	C. warmingii (NI)	6.8
" " " (ATCC)*	" " " (NI)	9.0

* Controls in which the cultures were not membrane-separated, but
separated by a piece of impervious rubber sheeting.

Dry Weight. The growth of the cultures was followed by dry
weight: optical density could not be used since sulfur globules
which accumulate within the cells interfere (9). Dry weight was
measured by filtering 10 ml samples and drying the filter membrane
plus culture in a desiccator. The filters used were Millipore
type GSWP 025 00, pore size 0.2 μm.

Inocula. Chromatium: 5 ml freshly prepared overnight culture. Desulforistella: 1 ml freshly prepared overnight culture.
Culture purity was checked frequently during growth, both micro-
scopically and by plating out.

Protein Determination. Protein was determined by the biuret
method (10) and expressed as % of dry weight. The standard was
bovine serum albumin (BDH).

Sulfur. These experiments were performed in 4 oz. Brockway
bottles containing Pfennig's medium at pH 9.0, with either sulfide
or sulfate as the sulfur sources. Parallel cultures were set up
as follows: 5 bottles C. warmingii (NI); 5 bottles Desulforistella sp.; 5 bottles mixed culture.
    The inocula totalled 1 ml in each case, the mixed culture
having 0.5 ml inocula of each of the two organisms. For each experiment, a bottle was taken of each culture and the sulfide and
dry weight measured. Sulfide was determined by the method of
Pachmayr as described by Trüper and Schlegel (11). The concentration of sulfide was measured by absorbance using a Beckman DB
spectrophotometer at 670 nm, and a standardized calibration curve.

Results

The results are given in Table I and Figures 3 to 9.
Table I presents the results of the protein determination.

Table I. Protein Determination Results

		Protein % of dry weight
Membrane-separated cultures		
Chromatium warmingii	pH 9.0	65 - 70
" " " " "	pH 6.8	62 - 68
Desulforistella sp.	pH 9.0	38
" " " " "	pH 6.8	32

Figure 3 shows the rate of growth of each organism in membrane-separated cultures at initial pH 9.0, and Figure 4 shows the changes in pH during growth of each organism. Figure 5 shows the rate of growth of the C. warmingii in pure culture, at two pH levels. Figures 6 and 7 present the results of growth measurement and sulfide determination in the bottle experiments using sulfide medium, and the final two figures, 8 and 9, give growth and sulfide during culture in sulfate medium.

Discussion

As is evident in Figure 3, the C. warmingii began to grow first, followed by very rapid growth of the Desulforistella. The final population of the C. warmingii was higher. The delay in onset of rapid growth by the sulphate reducer may be due to the need to accumulate sulfate being produced by the photosynthetic bacterium as it oxidized sulfide. It is probable that the sodium sulfide solution already contains a very small amount of sulfate, as Postgate (7) has suggested. Postgate also has reported that at least in Desulfovibrio an initial pH of 8.6 caused a more rapid growth of the organisms, but a reduced stationary population. It should be noted that the inoculum of Desulforistella was only one-fifth that of the C. warmingii, so the very rapid initial growth can hardly be attributed to inoculum size. At pH 9, at about 20 - 30 hours after the Desulforistella entered exponential phase the Chromatium cultures began their exponential phase.

Lower growth yields in this microbial system were observed at pH 6.8, and this may be attributable to the Desulforistalla, which did not grow well at the lower pH, and hence did not provide the needed sulfide for the Chromatium.

The change in pH of the membrane-separated cultures is shown in Figure 4. The medium as modified to pH 9.0 was not buffered, and the production of end products of the respiratory chain such as

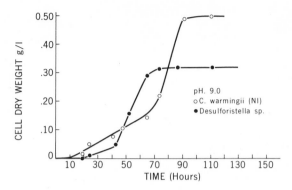

Figure 3. Membrane-separated culture, pH 9.0

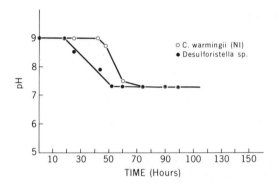

Figure 4. Membrane-separated culture, initial pH 9.0

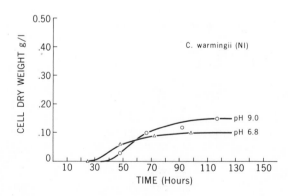

Figure 5. C. warmingii *(NI) growth at pH 6.8 and 9.0*

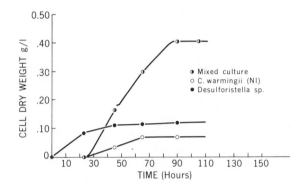

Figure 6.   Bottle cultures in sulfide medium, growth

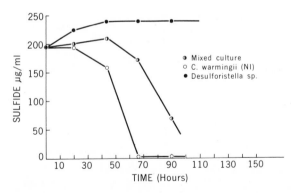

Figure 7.   Bottle cultures in sulfide medium, sulfide concentrations during growth

128

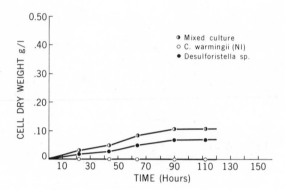

Figure 8.   Bottle cultures in sulfate medium, growth

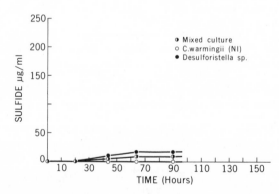

Figure 9.   Bottle cultures in sulfate medium, sulfide concentrations during growth

lactate caused the pH to fall to 7.4. The pH of the Desulforistella fell first at a time corresponding to very rapid growth (30
or 40 to 60 hours), and the later fall in pH of the Chromatium
also corresponded with the period of its most rapid growth (50 to
80 or 70 to 100 hours). At the end of growth, when both cultures
were in stationary phase, the pH was the same on both sides of the
membrane.

The growth of pure cultures of C. warmingii in this medium
was slow and the yields at stationary phase were much lower than
in the membrane-separated cultures, as seen in Figure 5. Their
growth was limited by the availability of sulfide. Trüper and
Schlegel (11) reported that the biomass yield of Chromatium could
be increased by the addition of sodium sulfide after the initial
supply was exhausted. At exhaustion of initial sulfide, their
yields were about 150 mg/l at pH 6.8, while in our case, at that
pH our yields were only 60 to 100 mg/l. Their study used C. okenii,
a species of larger bacteria than C. warmingii. They increased
their yields to 250 mg/l upon addition of a second supply of sulfide.

In the present study, yields of C. warmingii were somewhat
higher at pH 9.0 than at the lower pH, being 100 to 150 mg/l. In
these experiments, pH changes were much smaller in pure cultures
of C. warmingii than in the membrane-separated cultures, a result
associated with the lower yields.

The experiments with sulfide and sulfate media were designed
to relate changes of the sulfur compounds to bacterial growth, and
also to try to explain the increased yields at the higher pH. The
results are shown in Figures 6-9.

In sulfide medium, sulfide concentration increased during
Desulforistella growth, showing the reduction of sulfate, a probable contaminant of the sulfide. When this microorganism reached
stationary phase, no sulfate could be detected by barium precipitation. During Chromatium growth, sulfide decreased to almost
complete disappearance, at which point growth ceased. In the
mixed culture, however, there was an initial increase in sulfide
before its concentration began to fall. Mixed culture growth was
much more rapid. However, by about 90 hours, sulfide concentration
had fallen to a low level, and the mixed culture entered the stationary phase. The yield at stationary phase was about 400 mg/l,
whereas the pure cultures reached only 70 mg/l (C. warmingii) or
100 mg/l (Desulforistella). These pure culture yields obtained in
bottles are quite comparable to those obtained in pure culture
growth in the fermenters.

The results of growth in the sulfate media are shown in Figure
8, and the corresponding sulfide concentrations are shown in Figure
9. C. warmingii cannot utilize sulfate for growth (12, p. 37), as
found here. The C. warmingii did not grow. However, the Desulforistella did grow slowly to about 70 mg/l, while in the mixed
culture growth reached 110 mg/l. The mixed culture became pink in
colour, and the microscope revealed the presence of both the

Chromatium and the Desulforistella. Sulfide production closely
parallel the growth results. No sulfide was produced by the
Chromatium culture, while small amounts were produced by the
Desulforistella and in the mixed cultures.

From all these results it can be concluded that the interaction of C. warmingii and the Desulforistella sp. increased the
growth yields over pure cultures in both membrane-separated and
mixed cultures, and that this increase in yield can be explained
by the cycling of the sulfur between the two organisms, in the
manner designated by van Niel as a primitive cycle of matter.

The bacteria of this study were grown in mineral media. The
autotrophy of sulfate reducers has been questioned (7). Growth of
the Desulforistella on the mineral medium alone was not great.
However, in mixed culture or in membrane-separated culture, the
sulfate reducer grew very well, at times increasing 10-fold in a
24 hour period, a doubling time of about 7 hours. It is not clearly shown in this work whether the carbon for the growth came from
the carbonate or from organic compounds produced by the Chromatium.
Still it is beyond question that the microbial system is autotrophic.

As can be seen from the protein analysis data in Table I, C.
warmingii is rich in protein. Trüper and Schlegel (11) report a
protein content of 82.6% in their paper on C. okenii. They associate the high protein content with the chromatophore fraction of
the cells. Chromatium could be considered as a source of single
cell protein (13) and since it was grown on carbonate as a sole
source of carbon, its cost of production should be low compared
with single cell protein based on microbes grown on more expensive
carbon substrates. One can further consider a more complex system
containing a nitrogen fixer to further reduce substrate costs.

The low protein content of the Desulforistella raises the
question of the composition of the cells. Hvid-Hansen found a
"bituminous oil-like substance" in his cells, and he reflected that
finding in his specific name hydrocarbonoblastica. If Desulforistella is indeed capable of forming hydrocarbons, then the microbial
system we have been discussing may be a model for the formation of
petroleum under primitive cycle conditions, in which CO_2, as carbonate, is reduced to hydrocarbons. ZoBell (14) has discussed
several possible roles for microorganisms in petroleum formation,
but there is as yet little agreement on how microbes have participated in the process. The present work suggests one possibility,
namely that microbial photosynthesis is the primary process, and
that sulfate reducers such as we have been discussing, continue the
reduction of CO_2 to hydrocarbons.

The formation of high energy substances by fermentation is
well established in processes such as those which produce methane,
alcohol, or acetone and butanol. These processes require organic
substrates. Autotrophic microbial processes such as algal growth
have highly efficient photosystems and use low energy radiation,
that is the longer wavelengths. Part of their energy for CO_2

fixation and metabolism comes from the oxidation of sulfides. The energy balance in this microbial system is worthy of further study.

The above speculation about hydrocarbon formation suggests another advantage of membrane-separated culture. We can visualize production of high protein biomass in one fermenter, while high energy biomass is produced by interaction in another fermenter of such a microbial system.

Acknowledgements

This work was supported in part by the National Research Council of Canada. We are also grateful to Dr. Phillip Rueffel, Technical Manager, Saskatchewan Minerals Company, for supplies of microbial bloom from Durum Lake.

Abstract

This is a study of the anaerobic growth and interaction in membrane-separated cultures of a red photosynthetic sulfide oxidizer identified as Chromatium warmingii and a colourless, nonphotosynthetic sulfate reducer, tentatively placed with Desulforistella. The microorganisms were obtained from purple bloom in a sodium sulfate lake, Durum Lake, Saskatchewan. Each organism grew much better with interaction than in pure culture. The microbial system was autotrophic, carbonate being the sole source of carbon. Sulfide was utilized by the Chromatium and formed by the Desulforistella. The system fits van Niel's (1) description of a primitive cycle of matter. The Chromatium contained 62 to 70% protein, while the Desulforistella contained only 32 to 38% protein. These results suggest an autotrophic source of single cell protein, and also may have some bearing on the early formation of petroleum.

Literature Cited

1. van Niel, C.B., Methods in Enzymology (1971) 23, 3-28.
2. Pfennig, N., Annual Review of Microbiology (1967) 21, 285-324.
3. Smith, B.S., M.A.Sc. Thesis (1977), Department of Chemical Engineering and Applied Chemistry, University of Toronto, Toronto.
4. Arciero, G., Thesis (1977), Department of Chemical Engineering and Applied Chemistry, University of Toronto, Toronto, Canada.
5. Lapage, S.P., Shelton, J.E. and Mitchell, T.G., Methods in Microbiology, 3A, 119-120, Media Tables No. 76.
6. Butlin, K.R., Adams, M.E. and Thomas, M., J. Gen. Microbiol. (1948) 3, 46-59.
7. Postgate, J.R., J. Gen. Microbiol. (1950) 5, 714-724; Bacteriol. Rev. (1965) 29, 425-441.
8. Hvid-Hansen, N., Acta Pathologica Microbiol. Scand. (1951) 29, 314-334.
9. van Gemerden, H., Arch. Mikobiol. (1968) 64, 103-110.

10. Herbert, D., Phipps, P.J. and Strange, R.E., Methods in
     Microbiology (1971) 56, 244-248.
11. Trüper, H.G. and Schlegel, H.G., Antonie van Leeuwenhoek
     (1964) 30, 225-238.
12. Bergey's Manual of Determinative Microbiology, Eighth Edition,
     1974.
13. Protein-calorie Advisory Group of the United Nationls Organi-
     zation, PAG Bulletin (1976) 6(3), September.
14. ZoBell C.E., Science (1945) 102, 364-369.

RECEIVED July 25, 1978.

Potential of Biomass to Substitute for Petroleum in Canada

C. R. PHILLIPS, D. L. GRANATSTEIN, and M. A. WHEATLEY

Department of Chemical Engineering and Applied Chemistry, University of Toronto, Toronto, Ontario, Canada M5S 1A4

The attractiveness of production of liquid fuels from biomass lies in the renewable characteristics of biomass. As a consequence, the costs of an industry based on biomass conversion would be more or less predictable by inflation forecasting, and essentially independent of external political factors. With the incorporation of municipal solid waste as a biomass feedstock, such an industry also presents the opportunity of developing improved methods of recycling and waste disposal.

This paper is concerned with the potential for production of liquid fuels from biomass in Canada. To this end, the availability and cost of wood wastes, surplus roundwood, bush residues, energy plantation trees, and municipal solid wastes (mostly cellulosic) are assessed and promising thermal, chemical and biochemical conversion processes reviewed.

Liquid fuels have a high energy density, and the widest applicability of all fuel forms, but a low efficiency of conversion from biomass. It is therefore necessary to compare liquefaction with the more efficient processes of direct burning and gasification as alternative modes of use of the biomass.

During 1977, Canada consumed 658 million barrels of petroleum products (1). Products marked with an asterisk in Table I - motor gasoline and light, heavy and diesel fuel oil - account for 80% of the total production, and represent the main potential for biomass-based liquid fuel in Canada, almost 522 million barrels per annum.

Biomass Resources

Availability. It has been estimated that there are 450 million hectares of forest lands in Canada of which some 50% are presently productive and accessible. However, only 40-50% of the permissible annual yield is currently being used (2,3).

Table I

Refined Petroleum Products Canada –
January–December, 1977
Adapted from Reference (1)

Product	Barrels Produced
Propane and propane mixes	8,409,152
Butane and butane mixes	4,095,810
Petrochemical feedstocks	23,406,398
Naphtha specialties	3,707,347
Aviation gasoline	1,480,748
Motor gasoline*	225,762,377
Aviation turbo fuel	25,607,799
Kerosene, stove oil, tractor	26,080,239
Diesel fuel oil*	82,801,210
Light fuel oil (no. 2 and 3)*	87,621,595
Heavy fuel oil (no. 4, 5 and 6)*	125,612,433
Asphalt	18,423,926
Coke	4,874,929
Lube oil and grease	4,311,921
Still gas	21,869,082
Refinery losses	-9,122,863
Other products	3,728,415
Total production – all products	658,670,518

* These products (motor gasoline, diesel fuel oil, light fuel oil (no. 2 and 3) and heavy fuel oil (no. 4, 5 and 6)) together comprise about 80% of the total production

Unused wood residues as a by-product of current forest operations in Canada are estimated to be of the order of 0.14 billion cubic metres (4). Apart from what is presently being utilized, there exists an estimated annual roundwood surplus of some 0.2 billion cubic metres. Associated with this surplus would be a further 0.2 billion cubic metres of wood residues. If this wood were easily accessible and available at reasonable cost, it could be converted to methanol or fuel oil equivalent to about two-thirds of Canada's annual petroleum products production. In Ontario, Hall and Lambert (3) have estimated available quantities of surplus wood in several categories.

Agricultural wastes are disperse in character and of small total volume, and are best utilized on the farm. Municipal solid waste constitutes a disposal problem (12-20 million tonnes per year in Canada), thereby providing an incentive for its use. On the other hand, its heterogeneity (70-80% organic content) and availability in useful quantities only in large urban centres are disadvantages.

As well as surplus existing biomass, there is in Canada marginal farmland of some 30 million hectares (2). A small portion of this (about 1/2 million hectares in Ontario, for example) is presently known to be suitable for cultivation of hybrid poplar energy plantations.

Cost. Battelle Columbus (5) recently estimated the 1980 cost of readying timber residue, cull and dead trees for fuel conversion in the state of Vermont. The analysis considered procurement (stumping), harvesting, chipping and transportation over 40 kilometres, but omitted fertilization costs. Based on green wood (45% moisture, 10.9 GJ/green tonne), the wood cost was estimated as $16.40/green tonne.

In Canada, InterGroup Consulting Economists (6) estimated wood procurement costs in 62 forest zones across the country and generated cost data for the 20 zones having sufficient surplus roundwood to sustain a minimum 18,000 tonne per year methanol plant operation, based on a biomass recovery of 30%. The procurement costs shown in Table II represent the delivered chip cost, as in the Battelle study, but include capital and operating costs on an undiscounted cost basis. Unlike the Battelle study, the estimate is based on oven dry wood. The final results are very similar.

InterTechnology Corp. (7) analyzed the energy plantation concept, in which fast growing plant species are cultivated for biomass yield. For a plantation of hybrid poplar, total capital and operating costs in an unspecified U.S. location – from clearing the land to delivering wood chips – require a total revenue per oven dry tonne of $15.25 (in early 1975 dollars)(Table III). This assumes a 15% after-tax return on equity. The estimate is based on a yield of 20.2 oven dry tonnes per hectare. For a yield of 9

Table II

Zones of Surplus Forest Biomass and
Estimated Wood Procurement Costs
Adapted from Reference (6)

Province	Zone	Biomass Surplus (Thousand oven dry tonnes)	Procurement Costs (1976 dollars/ oven dry tonne)
B.C.	6	1,550	27.72
	8	2,210	39.78
Alta.	2,3	1,980	24.74
	4	2,100	25.26
	5	1,370	34.41
	6	2,180	33.85
	7	652	34.93
Sask.	1	828	31.06
	2,3	1,683	30.55
Man.	1	869	43.56
	3	707	26.88
Ont.	1,3	1,860 ⎫	40.93
	2,4	1,592 ⎬ *	39.85
	5,6	1,134 ⎭	35.79
P.Q.	2	2,960	41.40
	6,7	572	31.94
	8	1,908	37.55
	9	1,835	43.96
N.B.	1	620	26.37
Nfld.	1,2	382	28.84

* Of this, about 10^6 tonnes are considered to be realistically
available.

Table III

Breakdown of Capital and Operating Costs and
Estimate of Total Revenue Required per Year for
Deciduous Plant Material Grown on Energy Plantations
Adapted from Reference (7)

Basis: Annual productivity of plantation of 20.2 oven dry tonnes
per hectare per year (9 tons/acre year)

Annual plant material produced 9.37×10^5 oven dry tonnes
Number of plantation units   4 (46,200 hectares)
Average plant-material delivery distance   10 kilometres

COST ELEMENT:	Thousands $	Percent of total revenue required
Plantation Investment:		
1. Machinery and buildings	5,874	
2. Land clearing and preparation	5,514	
3. Total plant investment	11,388	
4. Interest during construction		
(a) Machinery and buildings	138	
(b) Land clearing and preparation	775	
5. Start-up	9,924	
6. Working capital	1,543	
7. Total capital investment	23,768	
Annual Operating Costs:		
8. Fuels	593	4.2%
9. Land rental	2,565	17.9
10. Payroll	3,337	23.3
11. Admin. and general overhead	790	5.5
12. Operating supplies	664	4.7
13. Repair parts	1,004	7.0
14. Local taxes and insurance	307	2.2
15. Total annual operating cost	9,260	64.8
Depreciation:		
16. Total Depreciation	2,895	20.2
Returns:		
17. Allowable gross return	1,566	11.0
18. Federal income tax	578	4.0
19. Total capital charges	5,039	15.0
Total revenue required	14,299	100.0%
Total revenue required per oven-dry tonne of plant material delivered	$15.25	

oven dry tonnes per hectare, revenue required rises to $34.39 per
oven dry tonne.

Sustained average yields from hybrid poplar NE-388 in Central
Pennsylvania, have reached 21.3 oven dry tonnes per hectare, with
0.37 square metres per plant, the first harvest after one year,
followed by five harvests at two-year intervals (7). Table IV
shows yield experience for hybrid cottonwood (from the poplar
family) in Canadian and American locations (8a-8f). Anderson and
Zsuffa (8) conducted growth tests on 35 different hybrid poplar
clones at Kemptville, Ontario. With 0.25 square metres per plant,
and harvesting after 2 years, they obtained yields of 4.9 to 19.3
oven dry tonnes per hectare year, with an average of 10.3 for the
30 clones that survived. Research is continuing into factors
required for maximum yield.

Municipal solid waste differs from wood biomass in that it
has a negative cost. In urban centres of Canada, waste is land-
filled at a cost of about $6 to $8 per tonne. While valuable
inorganics such as glass and aluminum can be recovered, the costs
of separation are high, and technologies are still developing.
With waste accumulating at the rate of about 1.6 kilograms per
capita per day, a need exists to reduce this bulk. However,
suitable quantities of municipal solid waste for fuel conversion
are available only in large urban centres.

Conversion Technology

The main conversion routes are to methanol (through synthesis
gas), to mixed oils by either thermal or chemical means, and to
ethanol by biochemical means. Alcohol products are compatible
only to a limited extent with existing end-use technology, for
example, the internal combustion gasoline engine and home heating
furnaces. A blend of 15% methanol in gasoline has been used in
Germany (9). Problems which require special attention include
phase separation, corrosion, driveability (surge, hesitation, cold
start, etc.). About 20% ethanol in gasoline has been used in
Brazil (10) where the climate is mild. The flashpoint of methanol
is 11°C which compares unfavourably with that of heating oil (38°C
minimum), thus indicating a serious safety hazard in use of
methanol in this application.

Mixed oil products have greater compatibility with existing
uses, although the substantial oxygen content of some of the products may pose certain refining and/or use problems. In the case
of substitute fuels not interchangeable simply with existing
fuels, the end-use costs - whether re-refining, efficiency
differences, distribution changes or equipment modifications -
should be included in the calculated comparative cost in dollars
per GJ.

Table IV

Yield Data for Hybrid Cottonwood
Adapted from Reference (8)

Species variety and location	Age at harvest (years)	Stems per ha	Oven dry tonnes per hectare per year	Reference
P. x euramericana clones I-214, Jacometti 78B and I-45/51				
S. Ontario	1	34,600	14.6 - 21.1[a]	8a
P. trichocarpa Washington	2	8,900 - 108,700	7.4 - 10.5[b]	8b
P. trichocarpa Washington	2	6,700 - 108,700	2.2 - 14.6	8c
P. trichocarpa Washington	2	unknown	2.2 - 13.5	8d
P. trichocarpa B.C. and Washington	2	6,700 - 108,700	8.1 - 11.7	8e
P. trichocarpa Washington	4	6,700 - 108,700	12.8 - 14.1	8f

[a] Oven-dry weight estimated by multiplying fresh weight by 0.49

[b] Oven-dry weight estimated by multiplying fresh weight by 0.46

Mixed Oil Products

Thermal. Although the Occidental Flash Pyrolysis system
(11) was developed for the organic content of solid waste, it
could readily be applied to a wood feed. Figure 1 (12) is a
flowsheet of the process based on 1360 tonnes per day of feed.
After separation of the inorganics, the dried (3% moisture)
organic fraction of the feed is finely shredded to 80% smaller
than 14 mesh (1200 micron). At this point, only 55 to 60% of the
original waste feed remains, and is carried into the pyrolysis
reactor by recycled product gas. Pyrolysis takes place at 500°C
and 100 kPa in the absence of air and catalyst. Pyrolysis occurs
at the reactor entrance where the feed is turbulently mixed with
hot ash from the previously combusted char (one of the products
of pyrolysis). The very short residence time produces the maximum proportion of liquid product, about 40% of the dry feed, or
about 1 barrel per tonne of waste. In addition, 20% char (19
MJ/kg) and 30% gas (14 MJ/m^3) are formed, as well as 10% water.
Due to its high oxygen content (about 30%), the product oil is
very viscous and of low energy content, only about 21–26 MJ/kg,
about one-half that of No. 6 fuel oil.

Tables V and VI (11) outline the estimated economics for a
910 and an 1820 tonne per day plant. The product revenue is
estimated as $13.23 per tonne of waste. For an 1820 tonne per
day plant, there is a net cost of $6.25 per tonne processed. On
the surface it would seem that an expensive process does no
better than landfilling the municipal solid waste at $6–$8 per
tonne. However, an important consideration is that the quantity
of unusable solids has been reduced to 16% by weight and less
than 5% by volume of the initial waste, a 20-fold decrease in
landfill space requirements.

Chemical. A chemical route is followed by two processes
which are similar, the U.S. Bureau of Mines (BuMines) waste
liquefaction process and the Worcester Polytechnic Institute
(WPI) hydrogenation process.

The BuMines process involves the reduction of cellulosics
with carbon monoxide, water and a sodium carbonate catalyst. A
formate is produced, which, it is thought, acts as a hydrogen
donor to reduce the feed. The following mechanism has been
proposed (13):

$$2CO + H_2O + Na_2CO_3 \rightarrow 2HCOONa + CO_2$$

$$2HCOONa \rightarrow H_2 + CO + Na_2CO_3$$

Figure 2 depicts the process equipment and Figure 3 is a flowsheet for a 2700 tonnes of prepared waste per day processing
plant (14). The prepared feed of 30% solids is slurried in water
and recycle oil and fed to the reactor at 350°C and 2 x 10^4 kPa.

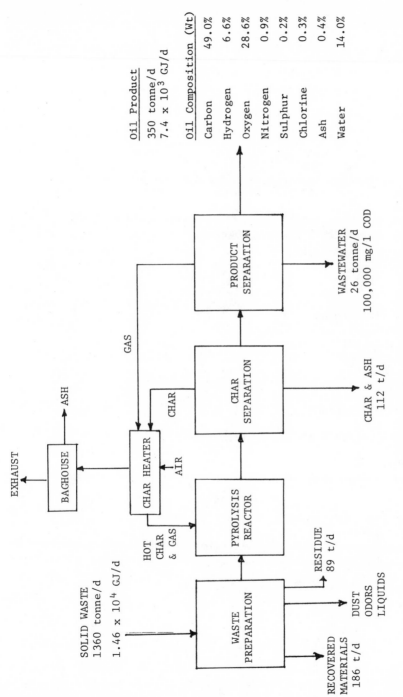

Figure 1. Occidental Research Corporation's flash pyrolysis system. Adapted from Ref. 12

Table V

Estimated Economics

Occidental Resource Recovery System Incl. Pyrolysis
Adapted from Reference (11)

Second quarter 1975 costs, geographically normalized

	910 tonne/d (1000 TPD)	1820 tonne/d (2000 TPD)
Capital Investment	$25.2 million	$37.9 million
Total Interest	26.1	39.2
Total Capital Cost	51.3	77.1
Capital Cost/tonne	$8.59	$6.46
Operating Cost/tonne	17.27	13.29
Revenue/tonne	13.23	13.23
Net Cost/tonne	$12.63	$6.52

Table VI

Product Revenues

Occidental Resource Recovery System Inc. Pyrolysis
Adapted from Reference (11)

Second quarter 1975 costs, geographically normalized

Product*	Net Price, $/tonne	Fraction of Refuse Recovered	Net Revenue, $/tonne Refuse
Ferrous Metal	45.38	0.0665	3.02
Glass Cullet	17.33	0.053	0.92
Aluminum	330.00	0.004	1.32

	Net Price, $/GJ	Energy Recovery, GJ/tonne Refuse	
Pyrolytic Oil	1.47	5.42	7.97

Total Net Revenue $13.23/tonne

* No revenue credit taken for product char at $27.50/tonne

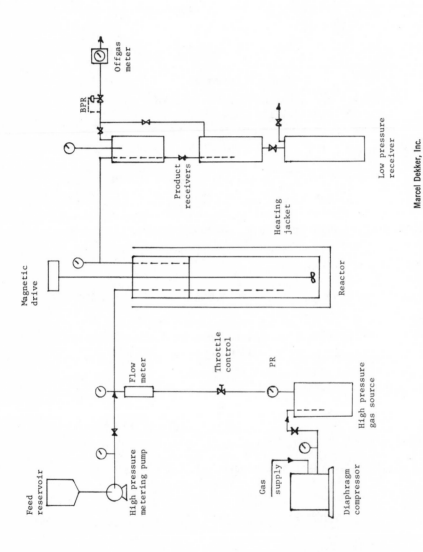

Marcel Dekker, Inc.

Figure 2. Continuous unit for converting waste to oil (14)

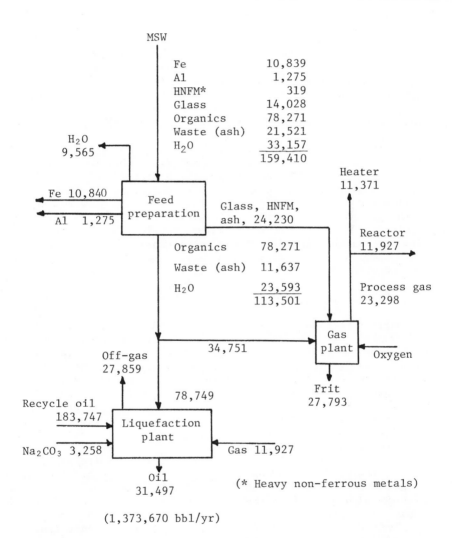

Figure 3. *Liquefaction of MSW (all streams kg/hr). Adapted from Ref. 14.*

Approximately 70% of the feed is utilized in the liquefaction
reaction, the other 30% being converted to synthesis gas for the
conversion and for process heat. The oil yield is 58 percent by
weight of organics, or roughly 200 kilograms per tonne of raw
feed. The product oil is viscous, due to its 15% oxygen content,
and has a heating value of about 32.5 MJ/kg. Tables VII and VIII
(14) outline estimated capital and operating costs in mid-1975
dollars for the liquefaction process. The break-even oil price
is calculated to be $9.53 per barrel or $1.61 per GJ.
    The WPI process is similar to the BuMines process, the major
difference being that WPI uses simple hydrogenation. Hydrogenation catalysts such as nickel hydroxide are used. Operating temperatures are around 450°C and operating pressures around 10,000
kPa. Based on pilot plant studies, a 50% slurry in recycle oil
would constitute the feed, and 2.37 barrels of oil per tonne of
wet organic waste would be expected as yield. Due to a low
oxygen content, usually less than 1%, the product oil is very
similar to heavy fuel oil, with a heat content of about 41.9 MJ
per kilogram. Table IX (15) is a cost versus capacity comparison
for the process (not stated, but believed to be in 1974 dollars)
based on a 32.7 tonne per day plant to service the town of Holden,
Mass. For larger capacity plants, this scheme appears to be the
most economically viable oil-producing process, due to its
expected superior yield and oil quality.

Methanol Products

    Whether methanol is produced from wood wastes, municipal
solid wastes, or a fossil fuel such as coal, the process consists
of the same two principal steps, the conversion of the feed to
synthesis gas (CO and H_2), followed by reaction to form methanol.
Gasification technology is still in the developmental stage,
whereas that for conversion to methanol is well developed, with
the last major innovation being use of a copper catalyst of high
activity to allow low pressure operation (commercialized in 1966
at the ICI plant in Billingham, U.K. (16)). The discussion here
is limited to the gasification stage.
    The Union Carbide Purox system (17) uses municipal solid
waste as the feed. Shredded ferrous-free refuse enters the top
of the conversion furnace and is contacted countercurrently with
hot combustion gases from the reaction occurring in the hearth.
As the solids proceed down the shaft, they are heated with progressively hotter gases. Initially, free water is vapourized.
Pyrolytic gasification of the organic portion of the refuse then
takes place, converting 50 to 60% by weight of the feed. The
pyrolytic material which consists of non-volatilizable carbonaceous material and inorganics is reacted in the hearth with pure
oxygen at 1650°C. The inorganics are melted to a fluid slag, and
hot gases are produced to carry out the pyrolysis. Gas produced
by the process is typically 21-23% H_2, 29-42% CO and 20-34% CO_2

Table VII

Estimated capital cost
(BuMines Liquefaction of waste)

Adapted from Reference (14)

	MSW ($000)
Feed Preparation	12,459
Drying	--
Reaction	50,604
Process Gas Production	13,306
Gas Purification	6,248
Offsites	28,916
Installed Plant Cost	111,533
Interest During Construction	22,307
Subtotal for Depreciation	133,840
Working Capital	13,384
Total Investment	147,224

Table VIII

Estimated operating cost
(BuMines liquefaction of waste)

Adapted from Reference (14)

	MSW ($)
Labour	795,000
Supervision	119,300
Fringe Benefits	274,300
Maintenance	3,346,000
Maintenance Supplies	2,230,700
General and Administrative	1,673,000
Insurance	1,115,300
Office Services	237,700
Utilities	1,114,000
· Depreciation	6,692,000
Taxes	1,673,000
Total Operating Costs	19,270,300
Byproduct Credits	
Iron	2,836,500
Aluminum	3,337,200
	6,173,700
Net Yearly Cost	13,096,600
Yearly Oil production (bbls)	1,373,600
Break Even Oil Price ($/bbl)	9.53
$/GJ	1.61

Table IX

Cost vs Capacity[a] (WPI)

Adapted from Reference (15)

Tonne/d	Capital Cost	Annual Operating Cost	Net Annual[b] Production bbl/year	Gross Cost of Production $/bbl
4.55	$341,000	$243,000	2000	121.50
9.1	517,000	265,000	4000	66.25
32.7	1,116,900	348,850	14000	24.77
91	2,060,000	498,000	40000	12.45
455	5,420,000	1,185,000	200000	5.92
910	8,200,000	1,900,000	400000	4.75
1820	12,400,000	3,200,000	800000	4.00

[a] based on 260 operating days/year (5 days/week)

[b] based on 2.37 bbl/tonne

(by volume) and energy produced is about 8 GJ/tonne refuse or the
equivalent of more than one barrel of oil. No cost figures are
available, but due to higher temperatures, costs may exceed those
of the Occidental Petroleum process.
    The Pulp and Paper Research Institute of Canada (PPRIC) pro-
cess (18) is based upon sawmill and pulpmill residues, mainly
bark. The process is similar to the Purox process in that heating
is countercurrent, and combustion is by pure oxygen. However, the
pyrolysis temperature is limited to 810°C, and product tar and
some product gas are used to provide heat for the reaction.
Typically, 89.5% of the dry feed is converted to gases, the
remainder being converted to tar and ash. The gas composition is
25.5% H_2, 27.7% CO and 21.2% CO_2 (by volume) and it is calculated
that energy produced is about 15.6 GJ/tonne dry feed. Based on a
yield of 32.5% of methanol by weight, and a delivered cost of
wood waste (in the Thunder Bay area) of $19.58/oven dry tonne,
PPRIC estimate a total product cost (for 910 tonnes methanol/day)
of $119.20 per tonne, or $5.98/GJ. (For a wood cost of $40 per
oven dry tonne and a conversion rate of 2.5 tonnes wood per tonne
of methanol, the methanol cost becomes approximately $10/GJ.)
Figures 4 and 5 (18) are schematics of the gasification and
methanol synthesis processes respectively.

Biochemical Processes

    There are two routes for the biochemical degradation of
cellulosic substances, (a) aerobic fungal or bacterial degradation
to produce glucose, followed by anaerobic yeast fermentation of
the glucose to form alcohol, and (b) anaerobic degradation, with
methane as the major product. Only the first process will be
dealt with here. A possible variation of this first process is
to use acid hydrolysis of wood to produce sugars for fermentation.
The Scholler process uses sulphuric acid, and the Bergius, hydro-
chloric. Disadvantages of acid hydrolysis include high capital
costs due to the corrosive nature of the acid, the need for high
steam pressures, and decomposition of the glucose by the acid
solution. Further, decomposition products such as methoxyfurfural
are toxic to the fermentation yeasts. Decomposition is partly
overcome in the Madison wood sugar process, where a continuous
flow system is used.
    Cellulose is found in nature in combination with various
other substances, the nature and composition of which depend on
the source and previous history of the sample. In most plants,
there are three major components: cellulose, hemicelluloses, and
lignin. Efficient utilization of all three components would
greatly help the economics of any scheme to obtain fuel from
biomass. Hemicelluloses, lignocellulose and lignin remaining
after enzymatic degradation of the cellulose in wood would
require chemical or thermal treatment – as distinct from biochemical – to produce a liquid fuel.

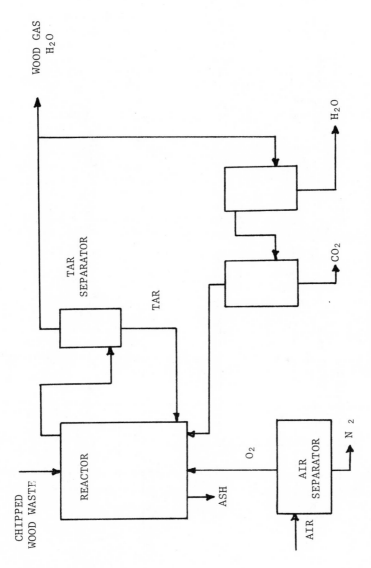

Figure 4. Wood pyrolysis (PPRIC) (18)

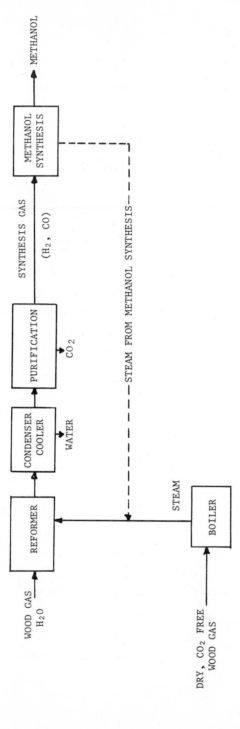

Figure 5. *Methanol synthesis from wood gas (PPRIC)* (18)

In considering the economics of fuel from biomass, it is necessary to consider not only alternative uses for the biomass, but also alternative sources and uses for glucose. Alternative carbon cycles are shown in Figure 6 (after Gaden (19)). Starch is a major competitor to cellulose for the production of glucose. Current technology shows that starch-derived glucose is more economical. In 1975, estimates in the U.S. for a 240 tonne/day glucose plant put capital costs for cellulose at $23 million, and for corn at $19.6 million. Corn gives a 99% glucose yield in a 30% solution, compared to a 76% glucose yield in a 4% solution in the case of cellulose. Use of a food grain for the production of ethanol is however a controversial issue. Production of ethanol from glucose must compete with the production of protein. Table X shows the conversion efficiency of sugar to various foods. It is clear that ethanol production must compete with growth of yeast for single cell protein and with the synthesis of sweet syrups. As developing countries become more advanced, demand increases for sweeteners, for soft drinks and foods. Single cell protein demand fluctuates because of its direct competition with soya and fish meal as animal feed. Single cell protein plants are of high capital cost, and require large scale operation.

Fermentation of sugars to alcohol is a well-established technology. Enzymatic hydrolysis of cellulose is more recent, and several groups are working on the process design and on economic evaluations. One of the major technological problems that is encountered is the necessity for pretreatment of the cellulose. Woody wastes are associated with lignin which is combined with the cellulose in such a way as to hinder access of the enzyme molecules to the cellulose. The other major problem is the degree of crystallinity of the cellulose itself, which must be reduced to a minimum to allow efficient enzyme saccharification.

Pretreatment may be by physical methods: grinding or milling, (all energy intensive processes), irradiation, high temperatures and pressures or combinations of these methods, or by chemical means, such as use of swelling agents like sodium hydroxide, ammonia, or phosphoric acid, or by delignification with gaseous sulphur dioxide under pressure, steaming or conventional pulping and bleaching methods.

Only a few fungi, such as *Trichoderma reesei, T. Lignorum* and *T. koningii, Chrysosporium lignorum* and *C. pruinosum, Penicillium funiculosum* and *P. iriensis,* and *Fusarium solani,* have been reported to produce high levels of cellulose degrading enzymes. Some bacteria, such as *Thermoactinomyces*, have also been studied. Cellulase enzymes are a complex interrelated set of enzymes, and for practical saccharifications a stable cell-free enzyme preparation with adequate levels of all the essential components is required.

A pilot study for large scale production of cellulase from *Trichoderma reesei* has been developed at the U.S. Army Natick Development Center (Fig. 7 (20)). In March 1977, enzyme produc-

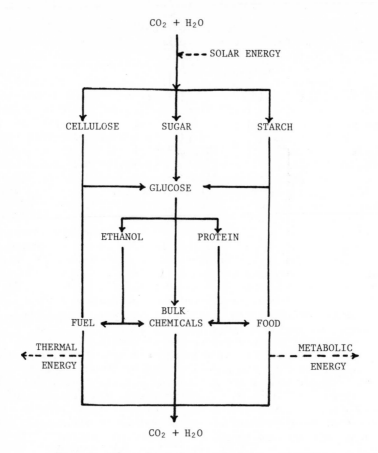

Figure 6.   The carbon balance. Adapted from Ref. 19.

Table X

Conversion of Sugar to Various Foods

Item	g Food/100 g Sugar
Beef	8 – 12
Poultry	10 – 15
Milk	15 – 20
Yeast	45 – 55
Sweet Syrup	90 – 100

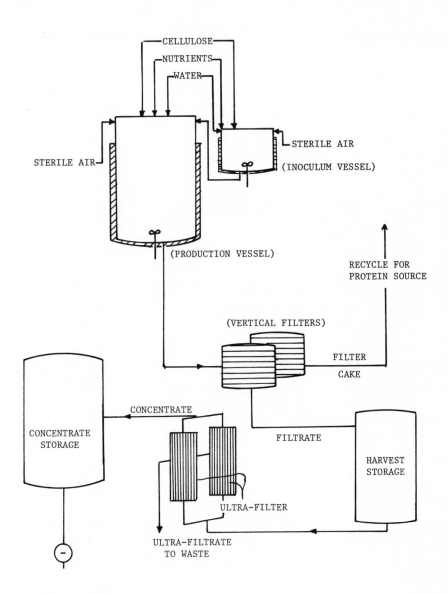

Figure 7. Pilot plant process for cellulase production. Adapted from Ref. 20.

tivity from this system was 42 International Units per hour. A
30 litre seed fermenter was used to provide inoculum for the
larger 400 litre production fermenter. A high degree of instrumentation is used to monitor and control the vessels during
operation. *Trichoderma reesei* was grown on 1% alpha cellulose
with basic salts added. Optimization of the process suggests
that maximum cellulase yield can be obtained 3 to 4 days after
inoculation. The fermenter liquor is then filtered to give a
filter cake of cell mass and unused cellulose (which can be
recycled). The filtrate is refiltered and concentrated for
hydrolysis.

Figure 8 is a schematic flow diagram for the hydrolysis of
waste newsprint. Most of the process design criteria and the
economic evaluations of the saccharification process have been
based on newsprint as substrate. Notable analyses are those of
Wilke and co-workers (21) and Humphrey (22). In the hydrolysis,
the substrate is first pretreated (milling), to make it more
accessible to the enzyme. Saccharification takes place in a
reaction vessel, where the substrate is contacted with the enzyme
solution from the fermentation vessel. Glucose solution is
separated from unreacted substrate at the outlet of the vessel
and the solution passes on to a concentration stage before the
sugar is used in the yeast fermentation to produce alcohol.

Most economic evaluations of the saccharification process
(21,22,23) conclude that at the present time the cost of production of the most favoured products (glucose, single cell protein,
ethanol) is higher than production from non-cellulosic sources.
Nyiri (24) made an economic evaluation of cellulose-based single
cell protein and ethanol production. He suggested that an
economical plant output is between 7^3 and 20^3 m^3/year, and,
depending on the size and complexity of the plant, estimated
capital costs between $6 and $12 million.

It can be concluded that the biochemical production of
liquid fuels from biomass is technologically feasible, but much
work is still needed to optimize the various aspects of the processes. The present day economic climate is not favourable for
production of these fuels from biomass by biochemical routes.

The Potential of Biomass to Provide Liquid Fuel in Canada

Biomass, the most useful form of which is wood, may be
burned directly, gasified or liquefied, with energy efficiency
decreasing in that order (Table XI). In addition, there are non-
energy uses, for example, for pulp, or in the case of juvenile
poplar leaves, potentially for cattle fodder. Clearly also,
there are alternative agricultural uses for the land used for
energy plantations.

In the case of direct burning, it can be argued that heavy
oil will be displaced for re-processing to produce transportation
fuel, which constitutes the highest value use of a liquid fuel.

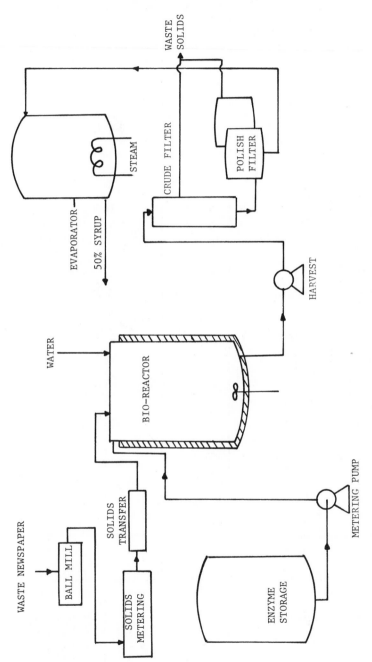

Figure 8. Pilot plant process for newspaper hydrolysis. Adapted from Ref. 20.

Table XI

Wood Usage[1,2] for Energy (Basis: $40/oven dry tonne)

Mode of Use	Efficiency (direct burning = 100)	Cost ($/GJ)	Applicability & Comments[3]	Action Required
Direct burning	100	2	About 2% of total petroleum consumption.[4] Displaces heavy oil for reprocessing.	Substitute for heavy oil.
Gasification	~60	5	Site specific (low energy density). Perhaps 1% of total petroleum consumption.	Some demonstration justified.
Liquefaction (Chemical routes)	~40	10-12[6,7]	Resource limited to less than 3% of total petroleum consumption based on current inventory of suitable energy plantation land.[5]	Some R & D justified (but wood cost is about one-half of liquid fuel cost)
Liquefaction (Biochemical routes)		>12[8]		

Footnotes to Table XI

1. An alternative wood use is pulp & paper.

2. An alternative energy plantation land use is agriculture.

3. Ontario case taken.

4. Estimated as 10% of heavy oil consumption.

5. Based on currently proven energy plantation area in Ontario of about 500,000 hectares and poplar yields of about 9 oven-dry tonnes per hectare per year (about 4 oven-dry tons per acre per year) plus about one million tonnes per year of wood residues. These feedstock quantities would produce about 2-3% of the total petroleum consumption. Some additional production is possible by expansion of the energy plantation area. (An additional one and one-half million hectares would increase the percentage liquid fuel production to about 8-10%.)

6. Cost is about $10-12/GJ for both methanol and synthetic gasoline for transportation usage. The additional conversion costs of methanol to gasoline by the Mobil process roughly balance the distribution and usage costs of methanol/gasoline blends.

7. Cost is 50-100% greater than gasoline for $30/bbl crude oil.

8. For ethanol.

Heavy oil may also be displaced by coal, natural gas (in relatively good supply in Canada until about the year 2000) and electricity. Interfuel substitution of heavy oil by solid fuels - coal
and wood - is estimated to have a potential to replace about 2%
of the liquid fuel consumption of Ontario (or about 10% of the
heavy oil consumption). Additional refining costs would be
incurred in converting the heavy oil to gasoline.
     Gasification represents the next most energy efficient use
of wood, and is estimated to have a potential to replace about 1%
of the total petroleum consumption, since the low energy density
of the gas requires site-specific use.
     Liquefaction is the least energy efficient mode of use of
wood. For the Ontario case, based on conversion of about 10^6
tonnes/year of wood and 500,000 hectares of poplar energy plantation (9 tonnes/hectare/year), liquefaction has the potential
to replace about 2-3% of the liquid petroleum consumption of
Ontario. Given the availability of a further 1.5×10^6 hectares
of suitable land, the potential would increase to about 8-10%.
It should be noted that large scale land use, whether for agriculture or for energy plantation purposes, raises environmental
issues.
     Direct burning is clearly less costly than gasification,
which, in turn, is much less costly than liquefaction (Table XI).
(The U.S. Dept. of Energy (25) has recently estimated similar
liquid fuel production costs.)
     Alcohols are more costly in end-use than hydrocarbon products
(for example, synthetic gasoline versus methanol) because distribution and equipment changes are usually required for alcohols.
The incremental cost of using methanol in gasoline is approximately equal to the incremental cost of converting the methanol
to gasoline by the Mobil process. Such end-use costs must of
course be included in overall fuel comparisons.
     The main competitive options for Canada for long-term liquid
fuel supply are shown in Table XII. (Short-term emergency supply
of liquid fuel for transportation is possible from natural gas,
either as methanol or as synthetic gasoline, albeit at substantial
cost. Direct use of compressed or liquefied gas in vehicles is
also possible.) As clearly indicated in Table XII, long-term non-
emergency priorities should be continued exploration for crude oil,
further development of the oil sands, and, possibly, liquefaction
of coal. Only after these options have been taken as far as is
appropriate should liquefaction of biomass be considered. This
may not be until near the turn of the century. In the interim,
however, direct burning of wood and wood gasification should be
pursued. Through this (and other) interfuel substitution, heavy
oil may be freed for conversion to transportation fuel. Although
these indirect routes of biomass utilization (direct burning and
gasification) are at present strongly preferred over liquefaction,
in the long term the scarcity of fuel supply from a renewable
resource base has attraction, and may represent an intangible

Table XII

Canadian Fossil Resources for Liquid Fuel

Fossil Fuel	Conversion Route	Action Required
Crude Oil at $20/bbl	→ Gasoline $4–5/GJ	Explore
Oil Sands	→ Gasoline $4/GJ	Develop
Coal	→ Methanol $5–6/GJ (for production in Western Canada) → Synthetic Gasoline $6–7/GJ	Hold Options Conduct R & D Perhaps Develop

benefit which can offset some of the higher cost of wood liquefaction. Continued research, development and demonstration is
therefore justified into reducing wood supply costs (at present
about 40-50% of the final liquid fuel cost) and into wood conversion technology.

Abstract

The potential of biomass to substitute for petroleum is
examined in terms of resource availability and cost, conversion
technology, and conversion and end-use costs. The most energy-
efficient and least costly mode of utilization of wood is direct
burning, followed by gasification, and, last, liquefaction.

Interfuel substitution of heavy oil by solid fuels such as
coal and wood and by natural gas and electricity has the potential
to replace about 2% of the liquid fuel consumption of Ontario.
The heavy oil thus freed becomes available for refining to a
transportation fuel.

Gasification of wood has a potential to replace about 1% of
the total petroleum consumption of Ontario.

Liquefaction of surplus wood and energy plantation poplar has
a potential to replace about 2-3% of Ontario's liquid fuel consumption (based on 500,000 hectares of presently known land
suitable for poplar plantations). If further suitable land areas
of 1.5×10^6 hectares were proven, this potential would increase
to about 8-10%.

Interfuel substitution via direct burning of wood is presently economical in certain cases, and wood gasification may be
attractive in site-specific applications. Wood liquefaction,
however, is not at present attractive.

The priorities in terms of liquid fuel production in Canada
should be: exploration for crude oil, further development of the
oil sands, and, perhaps, liquefaction of coal. Liquefaction of
wood should be pursued at a lower priority. Nevertheless, the
attraction of security of supply of liquid fuel from a renewable
resource does justify some research, development and demonstration
on wood production and wood liquefaction.

Literature Cited

1. "Refined Petroleum Products", Catalogue 45-004, 33(1),
    Statistics Canada, Ottawa (1978).
2. Middleton, P., Argue, R., Argue, R., Burrell, T., and
    Hathaway, G., "Canada's Renewable Energy Resources: An
    Assessment of Potential", Middleton Assoc., Toronto (1976).
3. Hall, R.J. and Lambert, L., "Resource Availability and Utili-
    zation of Forests for Energy", presented at "Alcohols as
    Alternative Fuels for Ontario" Symposium, Toronto, Ontario
    (19 November 1976).

4. Marshall, J.E., Petrick, G. and Chan, H., "A Look at the Economic Feasibility of Converting Wood into Liquid Fuel", Information Report E-X-25, Policy, Analysis and Program Development Branch, Canadian Forestry Service, Ottawa (1975).

5. Hall, E.H., Allen, C.M., Ball, D.A., Burch, J.E., Conkle, H.N., Lawhorn, W.T., Thomas, T.J. and Smithson, G.R. Jr., "Comparison of Fossil and Wood Fuels", Report No. EPA-600/2-76-056, U.S. Environmental Protection Agency, Office of Research and Development, Washington, D.C. (1976).

6. InterGroup Consulting Economists Ltd. (Winnipeg), "Economic Prefeasibility Study: Large-Scale Methanol Fuel Production from Surplus Canadian Forest Biomass, Part 1, Summary Report", Fisheries and Environment Canada, Environmental Management Service, Ottawa (1976).

7. Fraser, M.D., Henry, J-F., and Vail, C.W., "Design, Operation and Economics of the Energy Plantation", 371-395 in "Symposium Papers: Clean Fuels from Biomass, Sewage, Urban Refuse, Agricultural Wastes", Inst. of Gas Technol., Chicago (1976).

8. Anderson, H.W., Zsuffa, L., "Yield and Wood Quality of Hybrid Cottonwood Grown in Two-Year Rotation", Ontario Ministry of Natural Resources, Division of Forests, Forest Research Branch, Toronto (1975).

8a. Zsuffa, L. and Balatinecz, J.J., "Poplar Fibre Production in One-Year Rotation - The Potential of a New Concept", 14th Session Int. Poplar Comm., Bucharest (1971).

8b. Heilman, P.E., Peabody, D.V. Jr., DeBell, D.S. and Strand, R.F., Can. J. Forest Res., (1972) 2, 456.

8c. DeBell, D.S., Heilman, P.E., and Peabody, D.V., "Potential Production of Black Cottonwood and Red Alder at Dense Spacings in the Pacific Northwest", In "Abstr. 6th Forest Biol. Conf.", Tappi. Inst. Paper Chem., Appleton, Wis. (1972).

8d. Steinbeck, K., "Short Rotation Forestry: A Means of Combatting Shortages of Pulp Fibre", presented at Ann. Meeting Amer. Inst. Chem. Eng., New Orleans, La. (11-13 Mar. 1973).

8e. Smith, J.H.G., and DeBell, D.S., Forest Chron., (1973) 49, 31.

8f. Schmidt, F.L. and DeBell, D.S., "Wood Production and Kraft Pulping of Short-Rotation Hardwoods in the Pacific Northwest" in "IUFRO Biomass Studies Coll. Life Sci. Agr.", 507-516, Univ. Maine, Orano (1973).

9. Bernhardt, W., Konig, A., Lee, W. and Menard, H., "Economic Approaches to Utilize Alcohol Fuels in Automobiles", Inter. Symposium on Alcohol Fuel Technology, Wolfsburg, Germany (Nov. 1977).

10. Heitland, H., Czaschke, H.W. and Pinto, N., "Application of Alcohols from Biomass and their Alternatives as Motor Fuels in Brazil", ibid.

11. Preston, G.T., "Resource Recovery and Flash Pyrolysis of Municipal Refuse", in "Symposium Papers: Clean Fuels from

Biomass, Sewage, Urban Refuse, Agricultural Wastes, 89-114, Inst. of Gas Technology, Chicago (1976).

12. Gage, S.J. and Chapman, R.A., "Environmental Impact of Solid Waste and Biomass Conversion-to-Energy Processes", in "Symposium Papers: Clean Fuels from Biomass and Wastes", 465-482, Inst. of Gas Technology, Chicago (1977).

13. Appell, H.R., Fu, Y.C., Friedman, S., Yavorsky, P.M. and Wender, I., "Converting Organic Wastes to Oil: A Replenishable Energy Source", Bureau of Mines R.I. 7560, U.S. Dept. of Interior, Washington, D.C. (1971).

14. Del Bel, E, Friedman, S. and Yavorsky, P.M., "Economic Feasibility of the Conversion of Organic Waste to Fuel Oil and Pipeline Gas", in "Synthetic Fuels Processing: Comparative Economics", A.H. Pelofsky (Ed.), Chap. XX, 443-459, Marcel Dekker, Inc., New York (1977).

15. Kaufman, J.A. and Weiss, A.H., "Solid Waste Conversion: Cellulose Liquefaction", Report No. EPA-670/2-75-031, Nat. Environmental Research Center, Office of R & D, U.S. Environmental Protection Agency, Cincinnati (1975).

16. Moll, A.J. and Clark, C.R., "An Overview of Alcohol Production Routes", presented at "Alcohols as Alternative Fuels for Ontario" Symposium, Toronto, Ontario (19 November 1976).

17. Moses, C.T. and Rivero, J.R., "Design and Operation of the Purox System Demonstration Plant", from the 5th Nat. Congress on Waste Management Technology and Resource Recovery, Dallas, (7-9 Dec. 1976).

18. Azarniouch, M.K. and Thompson, K.M., "Alcohol from Cellulose-Production Technology", presented at "Alcohols as Alternative Fuels for Ontario" Symposium, Toronto, Ontario (19 Nov. 1976).

19. Gaden, E.L., "Biotechnology - An Old Solution to a New Problem", Chem. Eng. Div. Award Lecture, Amer. Soc. Eng. Ed. National Meeting (June 1974).

20. Spano, L.A., "Enzymatic Hydrolysis of Cellulosic Wastes to Fermentable Sugars for Alcohol Production", in "Symposium on Clean Fuels from Biomass, Sewage, Urban Refuse, Agricultural Wastes", 325-348, Inst. of Gas Technology, Chicago (1976).

21. Wilke, C.R., Stockar, V. and Yang, R.D., AIChE Symposium Series No. 158, (1976) 72, 104.

22. Humphrey, A.E., "Production of Food and Feed by Fermentation" AID Symposium on Utilization of Technology in Developing Nations, MIT, Cambridge (24-26 April 1974).

23. Allen, W.G. II, Biotechnol. and Bioeng. Symp. No. 6, 303, John Wiley & Sons, Inc., New York (1976).

24. Nyiri, L.K., AIChE Symposium Series No. 158 (1976), 72, 86.

25. "Alcohol Fuels Program Plan", Report No: DOE/US-0001/2, U.S. Department of Energy, Washington, D.C. (1978)

RECEIVED September 25, 1978.

Potential for Biomass Utilization in Canada

RALPH OVEREND

Renewable Energy Resources Branch, Energy, Mines and Resources, Canada

With the realisation that fossil fuels - crude oil, natural gas, tar sands and coals-are of finite extent, the use of renewable resources is advocated in many countries as a means of maintaining energy supplies in the light of declining oil reserves. Figure 1 shows the decline in crude oil production in Canada as well as the anticipated production of syncrudes from oil-sands over the next decades. Because of the decline of conventional production of oil, the transition to "non-conventional" sources of fuel has commenced and will accelerate as the next century approaches. Canada has announced its intention to exploit renewable resources; solar space and water heating and forest fuels are the subject of a recent government program intended to increase the proportion of energy derived from renewable sources from the present 3 1/2% to around 10% in the year 2000. In order to achieve this rapid growth it will be necessary to develop suitable technologies so that renewable energy will be able to contribute to what is essentially a world fueled by liquid and gaseous hydrocarbons. In the case of solar thermal applications, the key will be substitution of solar heat in applications and processes presently served by fossil fuels. For biomass, the key to large scale utilisation will be the conversion of photosynthetically produced material to forms and products compatible with the delivery and end use systems of the petroleum era. It is therefore necessary to analyse those resources and technologies available so as to identify those areas where there is a shortfall in the systems and technologies, before advocating any specific R&D strategy.

What is required is a form of resource and technology assessment. This could be summarized in a matrix in which the potential of each resource and technology can be quantified and charted. This matrix is shown in figure 2; the items in each column are illustrative and the lists are by no means exhaustive. The identification of R&D requirements then requires a techno-economic assessment of the possibility of different pathways through the matrix, and is the product of (Resource)* (Transport)* (Conversion Technology)* (Transport)* (End Use).

166

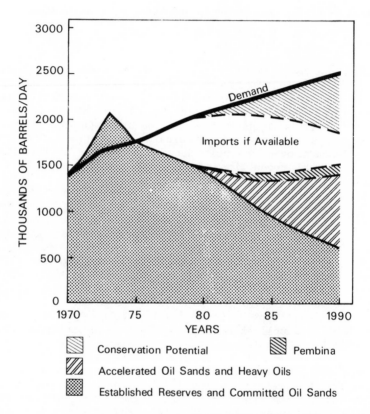

Department of Energy, Mines and Resources, Canada

Figure 1. *Canadian oil demand and availability, high price scenario 1970–1990*
(3)

RESOURCE	HARVEST & TRANSPORT	CONVERSION TECHNOLOGY	TRANSPORT	END USE
1- FOREST	ROAD	COMBUSTION eg: Steam Electricity	STEAM	PROCESS HEAT
MILL RESIDUE	RAIL		ELECTRICITY	ELECTRICITY
FOREST RESIDUE	BARGE	GASIFICATION eg: Low BTU Gas Synthesis Gas	PIPELINE	CHEMICAL SYNTHESIS
PRIMARY HARVEST	PIPELINE		TANKER	TRANSPORTATION
2- AGRICULTURE		PYROLYSIS eg: Char Oil	ROAD	FUELS eg: Methanol
FIELD RESIDUE			RAIL	
ANIMAL RESIDUE		HYDROGENATION eg: Oil SNG		
ENERGY CROPS				
3-AGRO-FORESTRY		FERMENTATION eg: CH_4 C_2H_5OH		
ENERGY PLANTATION				
4- MARICULTURE				
FRESH WATER				
SALT WATER				
1-Estuaries 2- Open Ocean				

Figure 2. Biomass technology and resource assessment chart

The pathways identified will not all satisfy the constraints that Canada's resources and needs identify, particularly in view of Canada's endowment of hydraulic resources,
nuclear potential, gas and coal reserves as well as non-
conventional oil resources such as oil sands and heavy oils.
These resources are not only constraints in the material and
economic sense but also in institutional ways. Examples of
such are: natural gas is a Canadian resource expected to be
available well into the 21st Century; hydraulic and nuclear
energy define a baseline in electricity costs of around 30
mill/kWh or $9/GJ (1977 $); methanol fuel for direct use in
automobiles will require replacement of a large quantity of the
downstream capital stock of the existing petroleum distribution
system. Again this is not an exhaustive list, but it serves to
emphasise that other criteria than the technical fact of being
able to transform biomass into any given product will probably
determine the eventual utilisation of this renewable resource.
To illustrate some of the R&D opportunities in Biomass, I will
discuss our present state of knowledge of some of the items in
the columns of the matrix in Figure 2.

Resources

Forestry. The total biomass productivity of Canada is not
known with certainty. Using data from Russian work for the
northern hemisphere (1), the total energy content of all the
carbon fixed in Canada by photosynthesis in one year is around
100 EJ (1 EJ = 10^{18} Joule = 0.948 Quad). This figure is an
extremely small fraction of the solar energy that falls on the
Canadian land mass in a year. The mean solar intensity of
$110W/m^2$ on Canada's 9.96 million square km corresponds to an
annual input of 34 000 EJ. While biomass offers the most
immediate opportunity in solar renewable energy, ultimately it
is the collection and transformation of solar energy in a more
direct fashion that will provide the majority of the renewable
energy supply in the distant future. At present, however, the
less than 0.5% biological efficiency of transformation of
solar-energy is compensated for by the inherent storage characteristics and high thermodynamic availability of the biomass
resource. Renewable energy in the form of hydraulic impoundment
and biomass contribute 0.8 EJ (2) and 0.35 EJ (3) respectively
to the Canadian primary energy supply (totalling 8.4 EJ in 1974)
and while this can be increased using known technology, the use
of solar and aeolian energy still requires considerable advances
in storage technology to be economically competitive.
Terrestrial biomass is of course dependent on a non renewable resource - the soil - for mechanical support and the supply
and transport of nutrients to the growing plant. The Canadian
total land area of 996,699,000 ha has the following land classification (4).

Table I

Land Type	Area/10⁶ha	% of Total
Water	81.006	8.1
Wildlife (tundra, muskeg, etc)	519.105	52.1
Agricultural	67.344	6.8
Urban and Other	6.199	0.6
Forest	323.045	32.4
	996.699	100.0

The most extensive land resource is the forest, covering 1/3 of the land mass. This is the foundation of the large forest industry in Canada: an industry employing 1 million people, 300 000 directly and 700 000 indirectly, and providing almost 20% of the value of the export trade. Data from FAO (5) sources show the statistics of the major forest countries marked in order of their forest inventories. (6)

Table II

Country	Population 10³ People	Forest Inventory 10⁶m³	Capita	Round Wood 10³m³	Production Capita
Brazil	109 730	74 315	677.2	163 995	1.49
USSR	255 038	73 250	287.2	387 600	1.52
USA	213 925	18 261	85.4	295 802	1.38
Canada	22 801	17 811	781.1	121 206	5.32
India	613 217	10 180	16.6	127 465	0.21
China	838 803	6 000	7.2	195 131	0.23

The forest opportunity in Canada for energy purposes in conjunction with the present industry is almost unmatched elsewhere in the world.

Thus, although approximately half of the wood harvested in the world is for fuel, it can be seen that only a few countries have sufficient forest resources on a per capita basis to satisfy a large proportion of their energy requirements. The energy demand in developed countries is presently around 10kW/capita while in the "Third world" it is around 500W/capita (7). Using a conversion of 1m³ wood/annum = 232W (based on the higher heating value of wood) it is easy to see why there is a developing firewood crisis (8). Indeed, all the countries in the table above would not be able to meet all of their needs from forest biomass.

The forest energy potential of Canada has been surveyed (9). The available material is essentially of 3 types. The lowest cost material is wood residue at the existing forest industry processing sites. While about half of the mill residue (bark, shavings and sawdust) is already utilised for energy, it is estimated that about 0.14EJ/annum could be further utilised in thermal conversion to steam and electricity in the forest industries that today use about 7.5% of all fuel oil and gas in such applications. The energy flow in the forest industry is summarised in figure 3.

Figure 4 is a generalised diagram of the distribution of biomass among tree components. Today only the bole of the tree is harvested from the forest; the tops, limbs and roots are left behind. With significant changes in harvesting technology - some of which are already in process - it will be possible to collect and utilise forest residue. For Canadian species, ex root, it is estimated that an increase of about 0.6 EJ over the present 1.0 EJ energy equivalent of round wood harvested could be obtained.

While the harvesting of trees for energy alone is likely to be too expensive, the unutilised forest of Canada could theoretically double the present wood harvest, though this would have to come from the area of the forest classified as tertiary below.

Table III

Forest Classification	Distance from Forest Industry Centres km	Area 10^6ha	%
Reserved	infinity	13.141	4.1
Primary	< 80	157.233	48.8
Secondary	< 80 < 120	19.849	6.0
Tertiary	< 120	132.822	41.1
		323.045	100.0

Thus a large infrastructure of roads and services or the improvement of water transport would be a prerequisite for such a development.

When harvesting of trees for energy can be tied to some other activity, such as intensive forest management, it may be possible to justify the recovery of some 0.37EJ in the form of currently non-commercial trees in primary and secondary forest regions. In summary, the energy potential of the forest system is:

Table IV

Source	10^6ODt/annum	Gross Energy Content /EJ	Cost $/GJ
Mill Residue	7.5	0.14	-0.2[a]
Forest Residue	31	0.58	0.8-1.2
Unutilized trees	20	0.37	1.0-1.5[b]
Tertiary Forest	52	0.97	2 ?

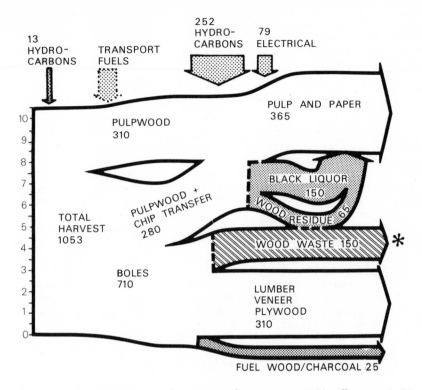

Figure 3. *Energy in the Canadian forest industry system 1974. All energy in PJ (10^{15} joule)*

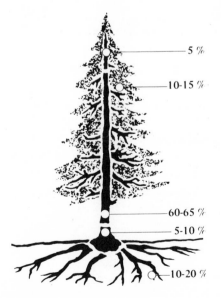

Figure 4. *Approximate proportions of tree biomass (northern species) in percentage of weight*

a)   currently incurs a disposal cost; its opportunity value
     relative to thermal coal at a forest industry site is
     0.8$/GJ or around $2/GJ relative to oil and gas.
b)   Will depend on costs being shared with forest management
     objectives.

Agriculture. The relatively small proportion of the
Canadian land mass suitable for agriculture is used to supply
the domestic and export food system. The role of Canada in the
provision of grain to other countries is not likely to change
in the foreseeable future, and the major agricultural energy
potential will be in residues of various kinds. Even these may
have higher opportunity values as emendation for soils, as
animal feed substitutes or substrates for single cell protein
production.
     Figure 5 after Leach (10) indicates the relative energy
efficiencies of current Canadian forestry, primitive and
western agricultural practice expressed as Energy Content of
Products/Energy Inputs for the products at the farm gate or
forest industry yard. Agricultural practice has moved to the
substitution of capital and energy for labour, along with the
use of high energy content inputs such as herbicides, fertilizers and irrigation.
     The energy content of Canadian agricultural production (11)
is set out below:

Table V

AGRICULTURAL PRODUCTION OF BIOMASS AS FOOD AND FEED

Product	Quantity 10^6 tonnes	Energy PJ
Cereal grains	33.3	579
Oilseeds	2.2	44.8
Forages	35.2	590
Pasture	43.1	627.7
Fruit, vegetables and potatoes	4.15	10.1
Dairy products	9.2	21.
Meat and poultry	2.0	21.6
Total of plant origin	117.95	1 851.6

Table VI

AVAILABLE CANADIAN BIOMASS PRODUCTION WASTES AND RESIDUES

Product	Quantity 10^6 tonnes	Energy PJ
Animal waste+	12.6	278.2
Crop residues*	16.9	246.4
Total	29.5	525.1

+   Animal waste is based on the assumption that cattle in the
    east are in pasture approximately 1/3 time and that cattle
    in the west are largely on range and therefore only 1/4 of
    the manure is available.

*   Crop residue averaged at 1/2 ton per acre. Straw is avail-
    able at over 1 ton per acre in the east but much less in the
    west.

Specialised forms of energy recovery from agricultural
residues are feasible, but are mitigated against in general by
the costs of collection. This reflects the inherent storage
characteristics of biomass--annuals such as grains can yield
about 1-5 t/ha of straw, a long rotation forest site can have as
much as 100-200 t/ha of residue and 100 t/ha of roundwood
boles. The difference in intensity and the lack of season-
ability of tree (perennial) harvesting compared with agriculture with 3rd and 4th quarter harvesting, results in a large
economic disbenefit to agriculture as an energy source.
    Animal residues do exist at some locations in large concen-
trations, such as at beef feed lots and swine and poultry
operations. The absolute energy content is not very large but
the emergence of anaerobic digestion technologies enables
environmental and social objectives to be met along with energy
production (12).
    Agro-forestry. There is a great deal of interest world wide
in the development of high yield perennial species for food,
fibre and fuel production. Canada is a participant with 5 other
countries in an International Energy Agency project on short
rotation forestry for energy purposes. Hybrid poplar species
have been developed (13) which yield 10-25 t/ha/annum. If these
can be "farmed" without moving the net energy benefit to far
into the region "crops" in figure 5, then the earlier estimates
of the energy available from forest resources could be an
underestimate by as much as a factor of 3-5.

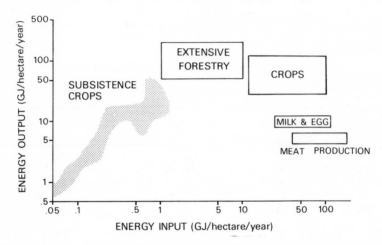

Figure 5. Energy inputs and outputs per unit of land for world farming systems

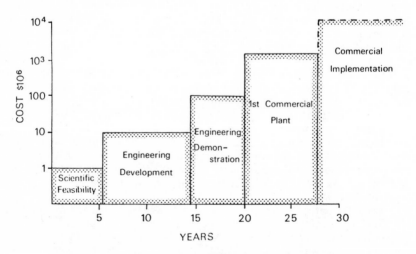

Figure 6. Costs of R, D & D with time for major energy projects

Harvesting and Transport of Biomass

There is a very large effort required to improve biomass harvesting and transportation. A survey of this field can be gained from the proceedings of the Biomass Energy Institute conference Forest and Field Fuels Symposium (14). The problem is one of materials handling where the material can be highly dispersed with areal densities from 5-250 ton/ha (green), depending on the species and rotation length. The material itself is usually around 50% moisture content and is awkwardly packaged by nature relative to man made collection systems. Drying, shredding, chipping, compaction and field conversions have all been tried with occasional success, but as yet this remains an area requiring extensive development.

Conversion Technologies

The biomass characteristics already described, namely 50% moisture content and heterogeneity, introduce many materials handling problems. The importance of scale should also be noted--the largest biomass processing plants are pulp mills of around 2000 ton/day of product or about 6000 t/day of dry feed stock. In electrical energy terms, the average Canadian pulp mill of 1200 t/day would correspond to an electrical station of 150 MW$_e$ (70-85% utilisation) or the production of about 1100 t/day of methanol. This apparent maximum size reflects the large harvesting area involved with its concomitant logistics problems. High yielding species such as the hybrid poplars referred to earlier could improve the scale from the same area to around 500 MW$_e$ equivalent or a factor of 3.

The R&D requirements for conversion technologies will reflect the end use demands of society, and it is here that there are great uncertainties. The costs of delivering new technologies and the time scales of developments are given in figure 6 (15).

The matching of biomass resources to the end needs of society is relatively easy to define over the next 5-7 years, since with the incentives already announced, the relatively low cost resource at mill sites can be utilized within the forest industry to back out of the equivalent of 70,000 bbl/day of crude oil, in applications mainly requiring steam or direct heat. The technology to do this at both large and small scales is already available, since in the light of the RD&D cycle defined in figure 6, the equipment is either fully commercial or the 1st commercial plant is under construction.

There are at present 2 or 3 low Btu gasifier installations at the engineering demonstration phase. Over the next few years these units are expected to become commercially available for deployment, and will be utilized both within the forest industry and at other sites where wood waste and plant residue are available.

The major R&D opportunity therefore lies with the resources other than mill waste which will nearly all be consumed at its point of generation within 7 years.

These other forest resources - unutilized trees from intensive forest management and the residue today left in the forest - could, if pressed to their maximum availability, contribute around 1 EJ to the energy supply. To do this will, however, require extensive end use product markets since the end use requirement of heat production in the forest industry will already be essentially satisfied by the industries' own residue. The conversion problem is therefore the transformation of biomass to energy intermediates such as electricity for transmission elsewhere, automobile fuels such as the much discussed methanol option, or into energy intensive tonnage chemicals such as ammonia and ethylene.

Methanol. If one takes as an illustration the many proposals (16,17) to use forest biomass to produce methanol as a motor fuel the R&D priorities and the barriers can be identified. The production route is:

biomass ⟶ synthesis gas ⟶ methanol

The R&D requirement exists solely in the production of synthesis gas from wood, since the catalytic production of methanol is well developed. The vertical shaft counter flow reactor developed by Union Carbide (18) for municipal waste could in theory be used to provide a synthesis gas from wood, and, as with Lurgi technology from coal, the RD&D phase is at the late engineering demonstration or indeed first commercial plant cycle. In general the economics of this are well understood and, given the poor economies of small scale 1100 t/day methanol plants, the product would cost around $8-10/GJ, 2/3 of which is capital, when prepared from a residue feedstock valued at around $1.50/GJ.

Other proposals could reduce the capital and operating costs: for example, a fluidised bed gasifier could increase the throughput for a given investment. The process of integrating external supplies of methane or electricity to produce methanol in hybrid renewable biomass/inexhaustible resources plants could use hydrogen generated from hydraulic or nuclear electricity or natural gas to supplement the carbon derived from sunlight. Both of these could reduce the capital cost and biomass requirements of methanol production significantly. It should however be recognised that these would then put the proposed system back to the engineering development stage or maybe even to the scientific feasibility stage of the RD&D cycle shown in figure 6.

The end use problem is similar in its consequences to the problem of when the methanol option can be delivered. Sweden

and Germany (19,20) have both demonstrated that with co-
solvents such as butanol, modern automobiles can run in
climates similar to Canada's on blends of 15-20% of methanol in
gasoline. The trend in vehicle development over the 5-6 year
minimum period to introduce methanol (with straightforward
production technology), could very well be to diesel and broad-
cut fueled vehicles. Since methanol and diesel are not miscible we are thrown further back into the development cycle if
we want to make significant inroads of biomass fuel into the
transportation sector.

The change in refineries and distribution systems required
for methanol incur not only direct economic costs, but also
attract resistance by the operating methods and social factors
of the petroleum system. There are other solutions, one of
which is the Mobil selective catalyst (21), which can prepare a
gasoline from methanol at very little energy penalty in an
integrated methanol/gasoline plant. This is in the engineering
demonstration phase today, and provided the economics are
improved, could be a superior option to using methanol either
directly or as a blend.

Figure 6 shows that the type of RD&D that is chosen is a
function of the time available before commercial application.
As well it indicates an obvious but under stated fact: any
RD&D activity can afford to have a large diversity at the
research or scientific feasibility stage, but much less at the
development or engineering development stage, and almost none
at the demonstration phase. Chemists will recognize this truth
for the development of polymers or drugs in industry just as
engineers and physicists do for the development of a nuclear
energy system such as CANDU (22).

Possible Paths

This article is not intended to be prescriptive and figure
7 represents only a personal view of the possible developments
in biomass energy. The timing of the use of large scale biomass reflects the shortfalls in "conventional" fuels forecast
in the WAES study (23).

The scientific feasibility stage of the RD&D cycle cannot
easily be directed - out of this curiosity phase it is hoped
that intelligent solutions will emerge in the transformation of
biomass to end-use products. Entropy considerations alone make
it offensive to think that with existing technologies we can
only make useful moities by breaking an elegant natural
structure down to synthesis gas and then chemically condensing
this to produce the desired molecule. The prediction here is
that biological techniques - mutant strains of bacteria, gene-
tic engineering, or whatever - will come along to tailor-make
the derived products by elegant scissions of natural structures. The large Canadian resource base should define this
direction even though at present it is very neglected.

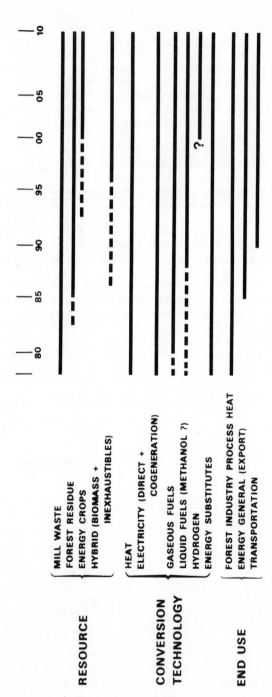

Figure 7.   Possible biomass energy utilization strategies

Biomass and the CO_2 problem

The increasing scale of human industrial activity has already increased the natural carbon dioxide content of the atmosphere from 260 vpm to 330 vpm, and it is increasing at around 1 vpm per year presently (24). There is little doubt that the root cause is anthropogenic and concern is now being expressed about the potential for fossil carbon to increase the level of CO_2 (an infra red absorber) to such an extent that a "greenhouse effect" will take over and change the climate of the world significantly, with unforeseeable consequences for food production and desertification.

Some authorities believe that biomass energy would serve to constrain the problem by setting up a cyclic system where photosynthetically fixed CO_2 is burnt to return as CO_2 to the atmosphere. Over a long time period this is correct, though with our present state of knowledge of the quantities and residence times of carbon in the major reservoirs it is not yet known whether or not terrestrial biomass is subtracting or contributing CO_2 to the atmosphere. Bolin (25) postulates that the destruction of the tropical forest is contributing to the increase as much as fossil fuel combustion. The residence time of carbon in the boreal forest is probably greater than 100 years, so that a rapid change to biomass energy from the forests would liberate carbon dioxide, increasing the atmospheric value even further before a steady state is reached. This assumes that the rate of replacement will be comparable to that of the initial biomass stock. Long periods without regeneration or, conversely, rapid regeneration with faster growing species, could alter the steady state value reached.

This problem could eventually severely constrain both the fossil and renewable carbon basis of future fuels, yet today a form of paralysis is the sole R&D activity discernible.

Conclusions

There is a large biomass resource available to Canada in the forest. On a per capita basis Canada could well have the greatest contribution of biomass energy of any country, yet the present per capita consumption of energy is already beyond the capability of the present biomass system to satisfy the total demand.

It follows that biomass will play a role complementary to other resources such as electricity from nuclear and hydraulic sources, as well as relatively inexhaustible supplies of natural gas, non-conventional oil and oil sands. The end use of these forms will be dictated by a combination of historic development and "technological inertia" such that substitution products: electricity, methanol, hydrogen, or tonnage chemicals like ammonia, will provide the major outlets for biomass carbon

before the next century after primary thermal uses have been
satisfied in the existing forest industries.

The role of competition in biomass for energy utilisation
should not be neglected. While biomass is a renewable form
derived from solar energy, it should be remembered that terres-
tial biomass is rooted in the soil, which is not renewable on a
time scale commensurate with human activities. I would place
poor stewardship of soil in the same risk category as the disposal of nuclear waste. Since the soil is finite, and society's
needs for fuel, fibres and food are proportional to a growing
population, biomass energy production may be hotly contested in
a hungry world and even the proposed agro-forestry projects may
be used primarily to provide protein and fibre, not energy.

The role of biomass in the natural carbon cycle is not well
understood, and in the light of predictions of a future atmospheric energy balance crisis caused by carbon dioxide accumulation, in turn the result of an exponential increase in the
consumption of carbon fuel, the apparent lack of concern by
scientists and policy makers is most troubling. Yet there is
no other single issue before us in energy supply which will
require action long before the worst effects of excess production will be apparent. The only satisfactory model is the
action taken by the R&D community with respect to the SST in
nitric oxide potential and chloro-halocarbon emissions, when it
was realised that the stratospheric ozone layer was vulnerable
to interference. Almost all other responses to "pollution"
have been after definitive effects have become apparent.

Literature Cited

1 Rodin, L.E., Bazilevich, N.I., Rozov, N.N., "Productivity of the World's Main Ecosystems," "Productivity of World Ecosystems: Proceedings of a Symposium," National Acedemy of Sciences, Washington, D.C., 1976, pp. 13-26.

2 Energy Policy Sector, Department of Energy, Mines and Resources, Report E1-77-5, "Electric Power in Canada 1976," Ottawa, Canada, 1976.

3 Overend, Ralph, "Energy and the Forest Industry (Estimates of Energy Flows)," Technical Note 5, Renewable Energy Resources Branch, Department of Energy, Mines and Resources, Ottawa, Canada, 1978.

4 Department of Fisheries and Environment, "Canada's Forests 1976," Ottawa, Canada, 1976.

5 Food and Agriculture Organization, "Production Yearbook 1975," Vol. 29, Rome, 1975.

6 Perrson, Reidar, "World Forest Resources: Review of the World's Forest Resources in the Early 1970's" Royal College of Forestry, Stockholm, 1974.

7 Hafele, W., "Energy Options Open to Mankind Beyond the Turn of the Century," International Institute for Applied Systems Analysis, Laxenburg, Austria, 1977.

8 Bene, J.G., Beall, H.W., Côté, A., "Trees, Food and People: Land Management in the Tropics," International Development Research Corporation, Ottawa, 1977.

9 Love, Peter, Overend, Ralph, "Tree Power: An Assessment of the Energy Potential of Forest Biomass in Canada," Report ER 78-1, Renewable Energy Resources Branch, Department of Energy, Mines and Resources, Ottawa, Canada, 1978.

10 Leach, Gerald, "Energy Accounting in Food Products," "9th International TNO Conference: The Energy Accounting of Materials, Products, Processes and Services," Rotterdam, Feb. 26 and 27, 1976, pp. 51-65.

11 Timbers, G.E., Downing, C.G.E., "Agricultural Biomass Wastes: Utilization Routes," Can. Agri. Eng. (1977) 19(2), 84-87.

12 Biomass Energy Institute, "Biogas Production from Animal Manure," Winnipeg, Manitoba, 1978.

13 Biomass Energy Institute, "Proceedings: Forest and Field Fuels Symposium," Winnipeg, Manitoba, Oct. 12 and 13, 1977.

14 Ibid.

15 Hill, G.R., "Critical Paths to Coal Utilization," Int. J. Energy Res. (1978) 1(4), 341-49, and personal communication, Ledwell, Dr. T.A., Renewable Energy Resources Branch, Department of Energy, Mines and Resources, 580 Booth St., Ottawa, Ontario, CANADA, K1A 0E4.

16 Intergroup Consulting Economists Ltd., "Economic Pre-Feasibility Study: Large Scale Methanol Fuel Production from Surplus Canadian Forest Biomass," Vols. I and II, Environment Canada, Ottawa, 1976.

17 Raphael Katzen Associates, "Chemicals from Wood Waste," Forest Products Laboratory, U.S. Forest Service, Madison, Wisconsin, 1975.

18 Schulz, Helmut W., "Energy from Municipal Refuse: A Comparison of Ten Processes," Prof. Eng. (1975) 45(11), 20-70.

19 Bern, L.A., Brandberg, A., "Methanol Supplied to the Swedish Motor Fuel Market," "International Symposium and Alcohol Fuel Technology: Methanol and Ethanol," Wolfsburg, Germany, Nov. 21-23, 1975, Vol. II.

20 Volkswagenwerk and the German Federal Ministry for Research and Technology (BMFT), "International Symposium on Alcohol Fuel Technology: Methanol and Ethanol," Vols. I-III, Wolfsburg, Germany, Nov. 21-23, 1975.

21 Meisel, S.L., McCullough, J.P., Lechthaler, C.H., Weisz, P.B., "Gasoline from Methanol in One Step," Chemtech (1976) 6(2), 86-89.

22 McIntyre, Hugh C., "Natural Uranium Heavy-Water Reactors," Sci. Am. (1975) 233(4), 17-27.

23 Workshop on Alternative Energy Strategies, "Energy: Global Prospects 1985-2000," McGraw-Hill, New York, 1977.

24 Woodwell, George M., "The Carbon Dioxide Question," Sci. Am. (1978) 238 (1), 34-43.

25 Bolin, Bert, "Changes of Land Biota and Their Importance for the Carbon Cycle," Science (1977) 196 (4290), 613-15.

RECEIVED September 25, 1978.

Material and Energy Balances in the Production of Ethanol from Wood

MORRIS WAYMAN, JAIRO H. LORA, and EDMUND GULBINAS

Department of Chemical Engineering and Applied Chemistry, University of Toronto, Toronto, Ontario, Canada M5S 1A4

Interest in renewable resources as raw materials for chemicals and energy has intensified in recent years as a result of anticipated shortages of petroleum and natural gas (1,2,3). A significant part of this effort has been devoted to the production of alcohols, particularly methanol and ethanol, from wood (4-8). Methanol is the main constituent of "wood alcohol", made for a great many years by the destructive distillation of wood, especially hardwoods (9). From a ton of hardwood, one could expect about 60 lb, that is 7.5 gallons of methanol, along with a variety of other chemicals. Wood alcohol is no longer made. The processes discussed in recent reports (4,5), are quite different, being based on wood gasification, followed by reactions similar to those employed in synthesis of methanol from natural gas. The expectation has been held out that 80 gallons of methanol could be obtained from a ton of wood, about 10 times the yield of the old wood alcohol. While there is general agreement that wood gas could probably be converted to methanol, to this date no one has done it, and there are no data, laboratory or otherwise, to indicate what problems might be encountered in the conversion, or what the economics might be. The recent suggestion that mixtures of wood gas and reformed natural gas may be a good source of methanol is most timely. Our calculations suggest that this may now be the cheapest route to methanol in many places.

The situation with regard to ethanol is much clearer: there is long industrial experience in the manufacture of ethanol from wood, by fermentation of the sugars in the waste effluents of pulp mills, or of the sugars made by wood hydrolysis (9). In the years following World War II, wood hydrolysis plants have been unable to compete economically with petroleum-based ethanol synthesis, mainly by hydration of ethylene, and they have been shut down in most countries. However, in the Soviet Union, we understand, there are still about 30 wood hydrolysis plants in operation (10). Many of these are used for fodder yeast production (11) but the wood sugars are also available for ethanol production. Recent market developments and technological advances have

served to stimulate interest in ethanol production from renewable
resources once again. The Brazilian national program for exten-
ding gasoline and even substituting for it by ethanol (12) has
been particularly noteworthy. The Brazilian experience has been
based on fermentation of the by-products of cane sugar manufacture,
and some ethanol has been produced there from cassava starch.
While ethanol from wood sugars has not been part of that program,
the successful substitution of ethanol for gasoline commercially
has raised the prospect of a large market outlet, and the rapidly
increasing price of gasoline here suggests that the economics, if
not favorable now, may become so in a foreseeable future.

Of the various technological advances which have been made in
wood treatment in recent years, the two most relevant to ethanol
production appear to be the autohydrolysis-extraction process for
hardwoods (13) and the enzymatic hydrolysis of cellulosic materials (14). By the autohydrolysis-extraction process, wood is
separated into its three main components, cellulose, hemicellulose
and lignin. The cellulose so produced is available for hydrolysis
by either acid or cellulase, while the hemicelluloses and lignin
are available for the co-production of chemicals or energy. By
autohydrolysis, that is steaming under carefully controlled conditions of time and temperature, the hemicelluloses are solubilized
and converted to sugars, furfural, acetic acid and other products,
while the lignin is so modified as to be extractable with caustic
soda under moderate conditions at atmospheric pressure, leaving
relatively pure cellulose undissolved. In this paper we report
our studies of this cellulosic residue, by hydrolysis and fermentation to ethanol.

The advances made in enzymatic hydrolysis of cellulosic
materials (14) are also of interest. This technology involves
only moderate temperature processes in simple equipment which pro-
mises to be of significantly lower capital cost than the pressure
equipment associated with conventional acid wood hydrolysis processes. All of these considerations combined to lead us to study
processes for ethanol production from wood, especially in an
effort to obtain data for material and energy balances, and possibly for the economics.

Pretreatment of the Wood

The particular wood species we chose for this study is aspen
(Populus tremuloides), which is plentiful in Canada and in the
northern U.S.A. The chemical composition we found to be glucan
53.4%, xylan 14.9%, total carbohydrate 79.0%, lignin 17.1% and
extractives 3.8%. We would expect total fermentable sugars of
about 56% in this sample of aspen in anhydro form (Timell has reported about 60% in another sample (15)) which upon hydrolysis
would yield about 1,250 lb wood sugars per ton of wood (dry
basis), from the stoichiometry. Theoretical conversion of this
sugar to ethanol would yield 640 lb or 81.1 gallons of anhydrous

ethanol, or 85 gallons of the 95% azeotrope.

The wood, in chip form, was treated by autohydrolysis (14). The chips, with initial moisture of about 50%, were heated with steam to 195°C and held at this temperature for 30 minutes. This time-temperature results in formation of volatiles such as furfural, water-soluble hemicellulose and alkali-soluble lignin. Typically about 12% of the wood substance was converted to volatiles, and about 12% of hemicelluloses was obtained in the water washings of the residue. About 75% of the wood substance remained as readily disintegrated fibrous pulp. Autohydrolysis was carried out in two different ways, continuous and batch. Continuous autohydrolysis was carried out in a reactor in the plant of Stake Technology Limited, Ottawa (16). The reactor was a horizontal tube about 10 inches in diameter and 8 feet long enclosing a helical screw conveyor. It was fed continuously by a plug-forming helix at one end, and discharged for a few seconds out of each minute at the other end. Steam was admitted at about 250 pounds per square inch pressure near the feed end, and measurements indicated that the temperature rose to the desired value instantaneously. The discharge through an orifice to atmospheric pressure sufficed to completely disintegrate the chips to a very fine pulp.

Batch autohydrolysis was carried out in 300 ml pressure vessels in a preheated silicone bath, with due allowance for temperature equilibration to achieve as nearly as possible 195°C for 30 minutes. Chip moisture was adjusted to 50% for the batch autohydrolysis.

For some experiments, the autohydrolysed pulp was subjected to hydrolysis, either acid or enzymatic, as formed, without washing or extraction. For other experiments, the autohydrolysed pulp after thorough water washing, was extracted with sodium hydroxide solution (20% NaOH on pulp, 70°C, 2 hours; 4% NaOH on pulp consumed) to remove the lignin. About 20 to 22% of the starting wood substance was recovered as lignin from the extract by acidification. The cellulosic residue, about 97% cellulose, was then about 50% of the starting wood substance. It was thoroughly washed and lightly acidified before hydrolysis.

As a result of these various pretreatments, the hydrolysis experiments were carried out on three different starting materials: autohydrolysed aspen wood made either in batch or in continuous equipment; and the cellulosic residue of the autohydrolysis-caustic extraction process.

Acid Hydrolysis

Acid hydrolysis was carried out in small (30 ml) pressure vessels in silicone oil baths at 190°C with dilute sulphuric acid. Upon completion, the vessels were cooled rapidly, the residue was thoroughly washed with hot water, dried and weighed. The liquor and wash waters were collected and the sugar content determined by

the anthrone method (17). Studies were made of the effect of
varying sulphuric acid concentrations, liquor to solids ratio, and
single stage compared with multi-stage hydrolysis.

   Single stage acid hydrolysis of the cellulosic residue of
autohydrolysis and extraction at 190°C with 0.5 to 4.0% H_2SO_4 on
the weight of the cellulose gave rather unsatisfactory results.
The maximum saccharification achieved was about 42% of that
theoretically possible. Degradation reactions caused the sugar
yield to drop. In multistage hydrolysis, the sugar was removed as
soon after formation as possible. Figure 1 shows the results of a
multi-stage hydrolysis with 2.0% H_2SO_4 on cellulosic residue by
heating for 20 minutes at 190°C, removing the liquor, washing the
residue with hot water, then repeating the hydrolysis with fresh
acid. The results show the considerable yield advantage of the
multistage acid hydrolysis, over 80% being obtained. This is
undoubtedly due to the removal of the sugar formed in each stage
of the acid hydrolysis, thereby preventing its destruction on
prolonged acid treatment. The same effect can, of course, be
achieved in other ways such as by pressure percolation or by
continuous hydrolysis in a properly designed reactor.

   The cellulosic residue used in these experiments is advantageous for acid hydrolysis compared to untreated wood. The
cellulosic residue is finely divided, and of higher density than
wood chips, 0.5 g/ml^3 compared to 0.36, and has a very low lignin
content. As a result, the output of fermentable sugars for a
given size of digester will be about twice that obtainable with
chips.

Enzyme Hydrolysis

   Cellulase enzyme, produced by Trichoderma reesei, a variant
of T. viridae, was kindly supplied by Dr. Mandels of the Natick
laboratory, and used for enzymatic hydrolysis at pH 4.8, 50°C of
autohydrolysed aspen wood obtained by continuous hydrolysis, and
cellulosic residue from the autohydrolysis-extraction process.
Sugar was determined by the dinitro salicyclic acid (DNSA) method
of Miller (18) as modified by the Natick cellulase laboratory
(19). The Natick laboratory had already tested autohydrolysed
wood samples prepared in the continuous reactor, with encouraging
results. Figure 2 illustrates our results. It shows a very
rapid initial reaction, slowing considerably after 24 hours. The
results in Figure 2 also show increasing saccharification when
higher amounts of cellulase are used. When 20% of the enzyme,
based on autohydrolysed wood weight, was used, the yield of sugars
was 77% of theory in 3 days. In another set of experiments using
10% by weight of cellulase, the action of the enzyme was continued
for 13 days. The yield after 3 days was 52%, in agreement with
the results shown in Figure 2, but the action of the enzyme continued, new sugar being formed at the rate of about 3% a day. At
13 days, when the experiment was discontinued, the yield was 90%

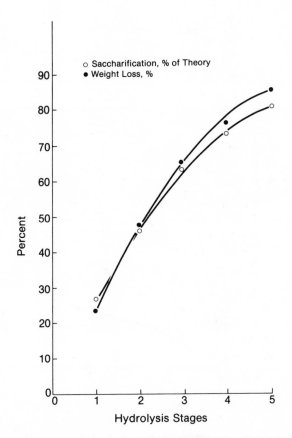

Figure 1. *Multistage hydrolysis of autohydrolyzed-extracted lignocellulosic residue, 2.0% H_2SO_4 on lignocellulosic residue; each stage 20 min at 190°C*

188   CHEMISTRY FOR ENERGY

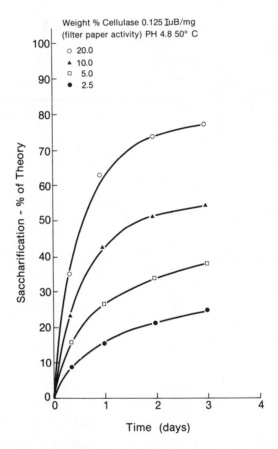

Figure 2.   Enzymatic hydrolysis of autohydrolyzed aspen wood (continuous auto-hydrolysis)

of theory, and the curve had not levelled off, suggesting that still higher yields were possible from the original addition of cellulase.

Such curves may be seen in Figure 3, which shows the results of cellulase hydrolysis of various substrates. In these experiments absorbent cotton was only slightly attacked by the enzyme, while the four other substrates were rapidly attacked in the first 24 hours. Thereafter, the cellulase continued its saccharification action, sugar yields of about 90% being obtained in 12 days from shredded filter paper. There were other very large differences observed. Autohydrolysed aspen prepared in the continuous reactor was almost as good a substrate as the filter paper, whereas the same process carried out in the batch reactors produced material highly resistant to enzymatic hydrolysis. The batch autohydrolysed aspen reached 30% saccharification in 1 day, then resisted further enzymatic attack. The difference is attributable to the physical form, since the discharge from the continuous digester involves an instantaneous change from about 200 pounds per square inch pressure to atmospheric pressure. This is accompanied by flashing off of steam and volatiles. The wood substance is thereby disrupted, and the cells separated and exposed. Upon extraction of the batch autohydrolysed material with NaOH, it became much more susceptible to enzymatic hydrolysis, sugar yields of about 65% being obtained in 12 days, but as is evident from Figure 3 it did not reach the rate of saccharification achieved with the continuously autohydrolysed wood.

Figure 4 shows the effect of multistage enzymatic hydrolysis of continuously autohydrolysed aspen wood. At the end of 2 days of cellulase hydrolysis at pH 4.8, 50°C with 10% of enzyme on wood, the sugar solution and enzyme were removed and the residue was washed with hot water. Fresh enzyme was then added and the hydrolysis continued for another day, when again sugars and enzyme were removed, and a new hydrolysis started. It is evident from the results of Figure 4 that by the end of day 5, 97.2% saccharification was obtained, and after 9 days the yield was 99.9%, the residue being a very finely divided black powder, most likely residual lignin.

The results of multi stage enzyme treatment suggests that further study of cellulase action is fully warranted. One hopeful area would be to decrease the hydrolysis time and enzyme concentration, and increase the number of stages. Another approach would be to remove the sugars by some other means, such as dialysis or fermentation (20).

Fermentation

Solutions of wood sugars were fermented anaerobically by ordinary Fleischmann's yeast (Saccharomyces cerevisiae). After 2 days the brews were tested for residual sugar. Also samples of the brew were distilled and the distillates analysed for ethanol

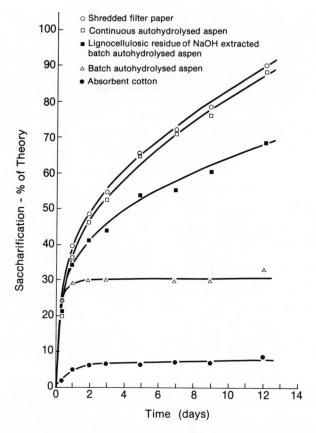

Figure 3. Enzymatic hydrolysis of various substrates by cellulase, 10.0% on sub-strate, pH 4.8, 50°C

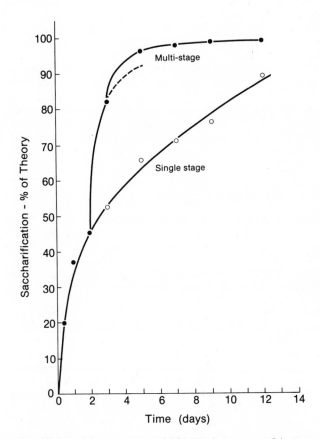

Figure 4. Enzymatic hydrolysis of autohydrolyzed aspen wood (continuous auto-hydrolysis), 10.0% cellulase on wood, pH 4.8, 50°C, single and multi-stage

192   CHEMISTRY FOR ENERGY

using gas chromatography. Complete fermentation of the sugars
occured when the sugars were derived from autohydrolysed-
extracted wood cellulosic residues, but only 72% of the sugars
were fermented in the 2 day period when they were obtained by
hydrolysis of autohydrolysed but not extracted pulp. The sugars
in the autohydrolysis liquors, that is the hemicellulose solutions,
were only 27% fermented in that period. This was not surprising
since most of the hemicellulose of aspen is xylan, a sugar not
fermentable by this yeast.

Alcohol recovery from the fermentation brews was less than
complete in most cases, which may be attributable to less than
ideal conditions. The best yields, 60 to 97% of theory, were
obtained with sugars obtained by hydrolysis of cellulosic residues
of the autohydrolysis-extraction process. Unextracted pulps, or
the hemicellulose solutions, gave poor ethanol formation, which
suggests inhibition. In the calculation of material and energy
balances which follows, we have assumed 95% yields of ethanol
from wood sugars, which is readily achieved in industrial practice and which we believe to be achievable with our wood sugars
as well.

Material and Energy Balances

The chart of Figure 5 shows material and energy balances for
ethanol production from aspen wood following autohydrolysis and
caustic extraction, including the results of acid hydrolysis of
the lignocellulosic residue, and the corresponding figures for
enzymatic hydrolysis. The numbers in both cases are based upon
the data obtained in each stage of processing as described above
in this report.

From the chart of Figure 5, we can expect to obtain from one
ton of aspen wood (dry basis), following autohydrolysis, caustic
extraction, acid hydrolysis and fermentation, 452.6 pounds or
57.3 gallons of 100% ethanol, or 58.4 gallons of the 95% ethanol-
water azeotrope. If enzyme hydrolysis is employed instead of acid
hydrolysis, somewhat more ethanol is obtained, 533.6 pounds or
67.6 gallons of 100% ethanol, corresponding to 68.9 gallons of 95%
ethanol. As was stated above, the theoretical expectation if all
stages gave 100% yields would be 649 lbs or 81.1 gallons. The
experimental results reported here suggest that we may expect
recovery of ethanol by these two processes at 70.7% and 83.4% of
this theoretical level. In addition we may expect to recover 426
lb of solid lignin and 246 lb of recoverable volatiles mainly
furfural and acetic and formic acids. The lignin so obtained may
be used as a solid fuel, or it may be converted to useful chemicals. The furfural in the volatiles is readily condensed and is
also a useful chemical. Acetic and formic acids are also articles
of commerce.

Energy balances shown on the charts of Figure 5 are based on
heat values re-determined here. These heat values are shown in
Table I.

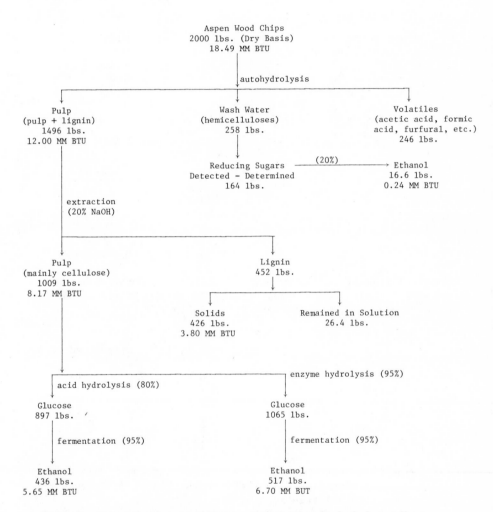

Figure 5. Material and energy balance. Acid and enzyme hydrolysis following autohydrolysis and caustic extraction.

TABLE I

Heat of Combustion Data

Substance	Heat of Combustion BTU/lb
aspen chips	9,246
autohydrolysed aspen	10,010
lignin	11,106
lignocellulosic pulp	8,098
ethanol (absolute)	12,968

Using these data we have the following gross energy recovery per ton of aspen wood:

	10^6 BTU	
	acid hydrolysis	enzyme hydrolysis
ethanol	5.88	6.93
lignin	3.80	3.80
volatiles	+	+
	9.68 +	10.73 +
	=52.35 +%	=58.03 +%

This gross energy recovery makes no allowance for processing energy, including autohydrolysis, acid hydrolysis and alcohol distillation. We can estimate it as follows:

	10^6 BTU	
	acid hydrolysis	enzyme hydrolysis
autohydrolysis	0.6	0.6
caustic extraction	0.1	0.1
hydrolysis	0.6	0.2
distillation	1.7	2.0
	3.0	2.9
net energy recovery	=37.1%	=42.3%

From these estimates it is apparent that the lignin recovered is capable of providing all of the energy required to operate the processes, with a little left over.

The chart of Figure 6 represents the results of acid hydrolysis and fermentation of autohydrolysed aspen. The yield of ethanol is slightly less than that obtained from autohydrolysed caustic extracted aspen being 436 lb or 55.2 gallons of 100% ethanol, or 56.3 gallons of 95% ethanol being obtained per ton of wood. The hydrolysis residue, a modified lignin, is obtained at the rate of 455 lb per ton of wood, and about 500 lb hemicelluloses and volatiles are also available for various uses. The gross

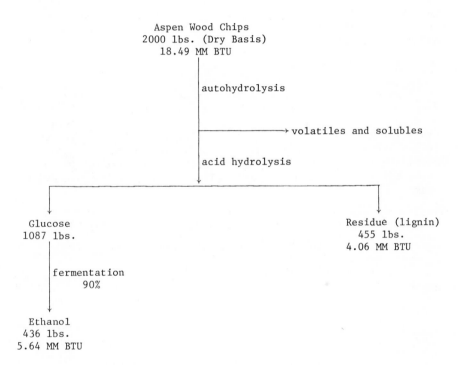

Figure 6.   Material and energy balance acid hydrolysis of autohydrolyzed aspen

energy recovery is 52.43%. Since there is no caustic extraction stage, a little energy will be saved in processing, and the net energy recovery is estimated at 36.8%. As before, there is enough lignin residue to fuel the process.

When autohydrolysed aspen was treated with cellulase enzyme, poor results were obtained. Based on a ton of dry wood, only about 300 lb of fermentable sugars were obtained, and the fermentation was very inefficient with low yields of ethanol. These results compare poorly with those shown in Figure 5 when autohydrolysed-caustic extracted aspen was treated with cellulase enzyme, and then fermented. It would appear from these results that the caustic extraction step for lignin removal would be necessary if the enzymatic hydrolysis process were to be adopted and advisable when acid hydrolysis is used.

Economics of the Two Processes

The data above enable us to make some rough estimates of costs associated with two processes: (i) ACID, that is aspen wood–autohydrolysis–caustic extraction–acid hydrolysis–fermentation–distillation; and (ii) ENZYME, that is aspen wood–autohydrolysis–caustic extraction–enzymatic hydrolysis–fermentation–distillation. For purposes of comparison, the product in both cases will be assumed to be 10 million gallons of 95% ethanol per year, a minimum economic size.

	Acid	Enzyme
Wood requirement, chips, dry tons/day	490	415
Capital Costs, $		
chip storage and distribution	1,000,000	850,000
autohydrolysis stage	4,000,000	3,500,000
caustic extraction stage	1,000,000	1,000,000
hydrolysis stage	4,000,000	2,000,000
fermentation	2,800,000	3,500,000
distillation	1,800,000	1,800,000
steam system	1,600,000	800,000
water system	150,000	400,000
material storage and handling	1,000,000	1,000,000
pollution control	150,000	150,000
buildings and land	3,500,000	3,500,000
total physical plant	21,000,000	18,500,000
engineering construction overhead, contingency	4,000,000	4,000,000
	25,000,000	22,500,000
working capital	3,000,000	3,000,000
	$28,000,000	$25,500,000

	Acid	Enzyme
Annual Operating Costs $		
Wood		
caustic soda	4,300,000	3,630,000
sulphuric acid	200,000	200,000
fermentation chemicals	600,000	1,000,000
packaging and distribution	1,000,000	1,000,000
total variable costs	6,100,000	5,830,000
utilities	100,000	100,000
labour	2,000,000	2,000,000
maintenance materials	200,000	200,000
other	200,000	200,000
total fixed costs	2,500,000	2,500,000
total annual operating costs	8,600,000	8,330,000
20% of investment	5,600,000	5,100,000
total annual costs	$14,200,000	$13,430,000
Price per gallon, including distribution and profit, but no tax	$1.42	$1.34

The above rough economics suggest a significant advantage for the enzymatic process, about 10% in capital costs and 6% in product "cost", or, since it includes an allowance for adequate return on investment, price. The price per gallon would make this ethanol competitive with industrial alcohol today, but it is too expensive to be considered for motor fuel at present gasoline prices.

The capital costs given are below those given by Mitre (21), or by Katzen (22), but are in line with those estimated by Robertson, Nickerson (23). It would take more data than are available to choose among these estimates. Our estimate is based on a flow sheet, inquiries of equipment costs and estimates based on other experience. Direct experience is, so far, not available. What is badly needed is a demonstration plant.

The distribution of factors which enter into the price may be summed up as follows:

variable operating costs	43%
capital related	39%
fixed operating costs	18%
	100%

The major factor which enters into variable operating costs is wood, here taken as $25.00 per dry ton, a reasonable current price for aspen chips in several locations in Canada. Under some special circumstances this can be reduced considerably: waste aspen chips, for example, would cost $18.00 per ton. If available in adequate quantity, such chips would cut the cost of ethanol by

about 10 to 12¢ per gallon. Government interest in industrial
employment might result in subsidies which could effectively cut
the capital cost in half. This would reduce the cost by 25 to 28¢
per gallon. Thus, it is conceivable that a properly located and
funded project could bring the price of this ethanol to $1.00.
This would represent a very good price for industrial ethanol. It
is still too high for gasoline, even taking into account expected
gasoline prices as now projected for the next decade, unless
unforseen shortages impose special advantages for this type of
motor fuel.

Discussion

     The present results suggest that the enzymatic hydrolysis
process is at least competitive with the acid hydrolysis process.
The main difficulty with it is the long time required for the
hydrolysis, compared with the acid process. The present work
suggested a mechanism for overcoming this problem, which is to
remove the sugars as formed, perhaps by combining the enzymatic
hydrolysis and the fermentation in one vessel at the same time.
In such a combined process, the removal of the sugars from the
solution as soon as they are formed by the fermentation would be
expected to increase both the yield and the rate of hydrolysis.
     This work also suggests other research and development dir-
ections needed to bring the price of ethanol down to an automotive
fuel level. We need a lower capital cost hydrolysis process which
can produce a concentrated sugar solution. We also need a fermentation process adaptable to concentrated sugar solutions to lower
alcohol purification costs. Finally we need to recover and include by-product values - lignin, furfural, acids, methanol, etc. -
in our income.
     We should also try to achieve economy of scale, which would
reduce unit cost considerably. Based on plantation hybrid poplar
(1), plants of 100 to 200 million gallons per year can be built -
10 to 20 times as large as the example used in these calculations.
This would reduce both capital and raw material costs. Perhaps we
need to be bolder in our approach to design of ethanol from biomass.

Summary and Conclusions

1. Yields of ethanol from aspen wood are 70.7% and 83.4% of theoretical where acid hydrolysis and enzymatic hydrolysis were
   employed. These were, respectively, 58.4 gallons and 68.9
   gallons of 95% ethanol per ton of aspen wood. In addition 426
   lb lignin with heat of combustion of 11,100 BTU/lb were
   obtained.
2. Gross energy recoveries (ethanol + lignin) by the two processes
   were 52.4% and 58.0%, respectively. Taking estimates of process energy into account, net energy recoveries were 36.1% and
   42.3%.

3. Multi-stage hydrolysis was beneficial, compared to single stage hydrolysis, for both acid and enzymatic hydrolysis. Sugar yields of 80% of theory were obtained in 5-stage acid hydro-lysis, and over 99% in 3-stage enzymatic hydrolysis. These results with enzyme hydrolysis suggest that constant removal of sugar as formed, by dialysis or by fermentation, may greatly improve present enzymatic hydrolysis procedures.
4. Wood pretreated by autohydrolysis and extraction is necessary for successful enzymatic hydrolysis, and advantageous for acid hydrolysis.
5. Economic estimates show a significant advantage in capital and operating costs for enzymatic hydrolysis ethanol. The invest-ment required for a plant to make 10 million gallons per year of aspen-based ethanol would be about $25 to $28 million. The price of the product, including a reasonable return on invest-ment, would be about $1.34 to $1.42 per gallon. About one quarter of the cost is wood cost, and another two-fifths is capital related. By a proper choice of location with good pro-ximity to cheap aspen chips, and by suitable funding arrange-ments, the price of 95% ethanol could be reduced to $1.00 per gallon. This would be a good price for industrial alcohol. It is too high to be considered as a gasoline substitute, unless unforseen shortages impose special advantages for this type of motor fuel.

Acknowledgements

We acknowledge with thanks the co-operation of Stake Technology, especially Mr. Robert Bender, in the continuous autohydrolysis, and of the Natick group under Dr. Leo Spano in providing cellulase. We have received financial assistance from the National Research Council of Canada and from the University of Toronto.

Abstract

Experimental production of ethanol from aspen wood gave yields of 70.7% or 83.4% of theory when acid hydrolysis or enzy-matic hydrolysis were used after autohydrolysis and extraction of lignin. These were, respectively, 58.4 and 68.9 gallons of 95% ethanol per ton of aspen wood (dry basis). In addition 426 lb of lignin with heat of combustion 11,100 BTU/lb were obtained per ton of wood. Gross energy recovery (ethanol + lignin) was 52.4 and 58.0% by the two processes, or allowing for processing energy, net energy recovery was 36.1 and 42.3% respectively. Multi stage hydrolysis was beneficial for both acid and enzymatic hydrolysis, 80% and over 99% of theoretical yields of sugar being obtained by the two processes. Economic estimates show a significant advantage in investment and operating costs for the enzymatic process. The price of 95% ethanol, including a reasonable return on investment by this process is estimated at $1.34/gallon. This would be a

200

good price for industrial ethanol, but would be quite high for
gasoline use under prevailing circumstance.

Literature Cited

1. Lora, J.H., M.A.Sc. Thesis, University of Toronto, 1976; Lora,
   J.H. and Wayman, M., Fast Growing Hybrid Poplar: A Renewable
   Source of Chemicals, Energy and Food, Forest Research Informa-
   tion Paper No. 102, Ontario Ministry of Natural Resources,
   Toronto, 1978; Wayman, M. and Lora, J.H., Wood-fired Electri-
   city Generation in Eastern Ontario, Royal Commission on
   Electric Power Planning, Toronto, Ontario, 1978 (including
   references to many recent symposia and studies).
2. Lewis, C.W., Fuels from Biomass – Energy Outlay Versus Energy
   Returns: A Critical Appraisal. Energy (1977) 2(3), 241–248.
3. Davies, D.S., The Changing Nature of Industrial Chemistry.
   Chemical and Engineering News (March, 1978) 22–27.
4. InterGroup Consulting Economists, Economic Pre-Feasibility
   Study: Large-Scale Methanol Fuel Production from Surplus Canadian Forest Biomass. Fisheries and Environment Canada, Ottawa,
   1976.
5. Mackay, D. and Sutherland, R., Methanol in Ontario. Ontario
   Ministry of Energy, Toronto, 1976; Mackay, D., Boocock, D.G.B.
   and Sutherland, R., The Production of Synthetic Liquid Fuels
   for Ontario. Ontario Ministry of Energy, Toronto, 1978.
6. Azarniouch, M.K. and Thompson, K.M., Alcohols from Cellulose –
   Production Technology, presented at Symposium: Canadian Society
   for Chemical Engineering and Canadian Society for Mechanical
   Engineering, Toronto, 1976.
7. Clark, D.S., Fowler, D.B., Whyte, R.B. and Wiens, J.K., Ethanol from Renewable Resources. The Canadian Wheat Board,
   Ottawa, 1971.
8. Robertson, Nickerson Limited, Saskatchewan Industrial Fermentation Complex, Regina, 1976.
9. Wenzl, H.F.N., "The Chemical Technology of Wood", Academic
   Press, New York, 1970.
10. Tokarev, B.I., Hydrolysis of Wood, Chapter XXIV in "The Chem-
    istry of Cellulose and Wood", by N.I. Nikitin. Translated
    from the Russian by Israel Program for Scientific Translations,
    Jerusalem, 1966.
11. Wayman, M., Food from Wood, Forest Commission Bulletin 56. Her
    Majesty's Stationery Office, London, 1976.
12. Linderman, R.L. and Rochioli, C., Ethanol in Brazil: A Brief
    Summary of the State of the Industry in 1977, The Seagram
    Lecture, the Second Joint Chemical Institute of Canada-
    American Chemical Society Conference, Montreal, 1977.
13. Gaden, E.L., Jr., Mandels, M.H., Reese, E.T. and Spano, L.A.,
    Enzymatic Conversion of Cellulosic Materials: Technology and
    Applications. Interscience-Wiley, New York, 1976.

14. Lora, J.H. and Wayman, M., Delignification of Hardwoods by Autohydrolysis and Extraction, Tappi (1978) 61(6) 47-50.
15. Timmel, T., Tappi (1957) 40(1) 30.
16. Stake Technology Limited, 20 Enterprise Avenue, Ottawa, Ontario, Canada, K2A 0A6.
17. Kohler, L.H., Differentiation of Carbohydrates by Anthrone Reaction Rate and Color Intensity, Anal. Chem. (1952) 24(10) 1576-1579.
18. Miller, G.L., Anal. Chem. (1959) 31, 426.
19. Mandels, M.H. and Sternberg, D., Recent Advances in Cellulase Technology. Cellulase Technology (1976) 54(4) 267-286. United States Army Natick Development Center. Production and Applications of Cellulase: Laboratory Procedures, Natick, Mass., 1974.
20. Myers, S.G., Ethanolic Fermentation During Enzymatic Hydrolysis of Cellulose. Second Pacific Chemical Engineering Congress, Denver, Colorado, 1977.
21. Blake, D. and Salo, D., Solar Related Technologies. Vol. IX. Biomass Fuell Production and Conversion Systems, The Mitre Corporation, McLean, Virginia, 1977.
22. Hokanson, A.E. and Katzen, R., Chemicals from Wood Waste. Ralph Katzen Associates, Cincinnati, Ohio, 1977.
23. Saskatchewan Industrial Fermentation Complex. Robertson, Nickerson Group Associates, Ottawa, 1976.

RECEIVED July 24, 1978.

14

Photochemical Aspects of Solar Energy Conversion and Storage

JAMES R. BOLTON

Photochemistry Unit, Chemistry Department, University of Western Ontario, London, Ontario, Canada N6A 5B7

Most of the current and proposed applications of solar energy involve its conversion to heat for space and water heating or to drive a Carnot engine to produce mechanical work or electricity. There are, however, some applications of solar energy which involve its conversion directly into electricity or to be stored as chemical energy without any thermal step in the process. These applications are quantum processes in that solar photons are employed to drive photophysical and photochemical processes. In this article, I will define qualitatively and quantitatively the thermodynamic and kinetic limits on the photochemical conversion and storage of solar energy as it is received on the earth's surface, evaluate a number of possible reactions with particular emphasis on the generation of solar fuels such as hydrogen from water and the generation of electricity.

A. General Requirements on the Photochemical Reaction

Many authors have considered the general requirements for useful solar photochemical reactions (1, 2, 3, 4, 5). In summary, they are:

1. The photochemical reaction must be endergonic.

2. The process must be cyclic.

3. Side reactions leading to the irreversible degradation of the photochemical reactants must be totally absent.

4. The reaction should be capable of operating over a wide bandwidth of the visible and ultraviolet portions of the solar spectrum with a threshold wavelength well into the red or near infrared.

5. The quantum yield for the photochemical reaction should be as high as possible.

In addition, there are some requirements which apply particularly to chemical energy storage reactions:

6. The back-reaction must be extremely slow under ambient conditions to permit long-term storage, but should proceed rapidly under special controlled catalytic conditions or elevated temperatures so as to release the stored energy when needed.

7. The product(s) of the photochemical reaction should be easy to store and transport.

8. The reagents and any container material should be cheap and non-toxic and the reaction should be unaffected by oxygen.

At present, the only photochemical storage system which satisfies nearly all of these conditions is the reaction of photosynthesis; however, certain photovoltaic cells such as the silicon and GaAs cells, satisfy many of the requirements for direct conversion to electricity. In addition, there are several possible systems which have a potential to satisfy most of the requirements and will be considered in this article.

B. Solar Energy Available at the Band Gap Wavelength

Direct conversion systems are threshold devices, that is, there is a minimum photon energy which can initiate the photochemical reaction. This is called the band-gap energy E_g with a corresponding band-gap wavelength λ_g. E_g usually corresponds to the 0-0 transition to the lowest excited singlet state of the absorber (see Fig. 1). Hence, it is important to know what fraction of the incident solar power is available at the band-gap energy.

If $N_s(\lambda)$ is the incident solar photon flux in the wavelength band from λ to $\lambda + d\lambda$ (in photons m^{-2} s^{-1} nm^{-1}) and $\alpha(\lambda)$ is the absorption coefficient of the absorber in the same band, then the absorbed flux of photons with $\lambda \leq \lambda_g$ is given by

$$J_e = \int_0^{\lambda_g} N_s(\lambda)\alpha(\lambda)\,d\lambda \qquad (1)$$

The available solar power E (Wm^{-2}) at E_g is thus

$$E = J_e \cdot \frac{hc}{\lambda_g} \qquad (2)$$

Then the fraction η_E of the incident solar power available to initiate photochemistry is

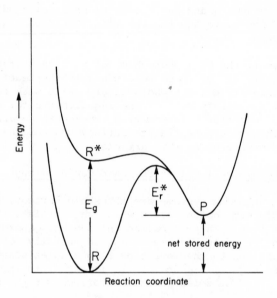

Figure 1.   Energy profile for a general endergonic photochemical reaction R → P. E_g is the minimum energy gap between the lowest vibrational levels of the excited state R* and the ground state R of the absorber. E_r* is the activation energy for the back reaction P → R.

$$\eta_E = \frac{E}{\int_0^\infty E(\lambda)\,d\lambda} \tag{3}$$

where the denominator in Eq. (3) is the total incident solar power.

As an approximation to the solar distribution at the earth's surface, we will take Boer's (T/S) distribution for AM 1.2 which is for a bright sunny summer day near noon (6). (AM stands for air mass. AM 0 would correspond to incident solar power outside the earth's atmosphere. AM 1 would correspond to incident solar power at sea level with the sun at the zenith.) If we make the ideal assumption that $\alpha(\lambda) = 1$ for $\lambda \leq \lambda_g$ and $\alpha(\lambda) = 0$ for $\lambda < \lambda_g$, then we can calculate an ideal value of η_E as a function of λ_g. This is shown as curve E in Fig. 2. η_E has a maximum value of 47% at 1110 nm for AM 1.2 but the maximum is very broad in that η_E is > 45% between 800 nm and 1300 nm. It is a common misconception among photochemists that η_E represents the fraction of solar power that can be converted to electricity or chemical energy. The next section will attempt to show the fallacy of this view.

C.   Thermodynamic Limits on the Conversion of Light Energy to Work

Many authors have treated the problem of thermodynamic limits on the conversion of light to work (e.g., electricity or chemical free energy) (7-12); however, Ross and Hsiao (13) have recently published a particularly lucid treatment which I will briefly summarize here.

Consider a dilute solution of a dye D in equilibrium with a black box at temperature T_L as shown in Fig. 3a. The black-body radiation will cause a very small but finite fraction of the dye molecules to be in the excited state D*. Let x_L be the mole fraction of D* molecules at equilibrium. Since the system is completely at equilibrium, the chemical potentials of ground and excited states must be equal, that is

$$\mu_D^e = \mu_{D*}^e = \mu_{D*}^o + RT \ln x_L \tag{4}$$

Now consider the situation in Fig. 3b with an external light beam irradiating the solution such that D is being excited to D* by the light. If D* has a sufficient lifetime (> 1 ps) such that a Boltzmann distribution is established among the vibrational levels of D*, then D* can be considered to be a separate chemical species which has come to thermal equilibrium with the black box still at the temperature T_L. Let x_H be the new mole fraction of D* where $x_H \gg x_L$. If we assume that the light beam is not so strong that the ground state would be significantly depleted (i.e. the mole fraction of the ground state is essentially unity with or without the light beam), then

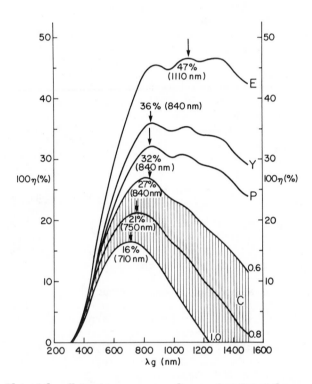

Figure 2. Plots of the efficiencies η_E, η_Y, η_P, and η_C as a function of the wavelength λ_g corresponding to the band gap E_g. The distributions have been calculated for AM 1.2 solar radiation (taken from distribution T/S of Ref. 6). Curves, E, Y, P, and C are plots of η_E, η_Y, η_P, and η_C as defined in Equations 3, 8, 12, and 16, respectively. η_C has been calculated for 0.6, 0.8, and 1.0 eV energy loss, respectively, as indicated on the figure.

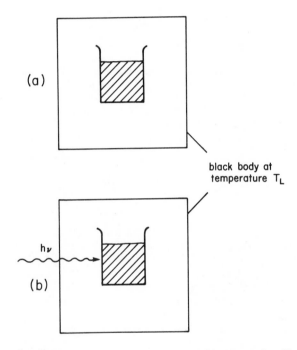

Figure 3. *(a) A dilute solution of a dye D in equilibrium with a black body at temperature* T_L *but with no external radiation; (b) same system as in (a) but with an external light beam irradiating the dye solution such that D is being excited to* D^* *by the light*

$$\mu_D = \mu_D^e$$

However, with the light beam the chemical potential of D* will increase and now

$$\mu_{D*} = \mu_{D*}^o + RT \ln x_H \tag{5}$$

Thus, the chemical potential difference between D and D* which can be generated by the light beam is

$$\mu = \mu_{D*} - \mu_{D*}^e = RT \ln (x_H/x_L) \tag{6}$$

Since μ increases as the population of D* increases, for a given light intensity μ will have its maximum value μ_{max} when the only route for decay of D* is via radiative decay (i.e. $\phi_{fluorescence} = 1.0$). Clearly μ depends on the light intensity and goes to zero as the light intensity goes to zero ($x_H \rightarrow x_L$).

Since μ_{max} represents the maximum thermodynamic potential, the maximum power yield is then

$$Y = J_e \cdot \mu_{max} \tag{7}$$

and the maximum thermodynamic efficiency is

$$\eta_Y = \frac{Y}{\int_0^\infty E(\lambda) \, d\lambda} \tag{8}$$

η_Y is plotted as curve Y in Fig. 2.

Ross and Hsiao (13) have derived the following expression for μ_{max}

$$\mu_{max} \doteq \frac{N_o hc}{\lambda_g} + RT \ln J_e - RT \ln [\frac{8\pi n^2 kT}{h\lambda_g^2}] \tag{9}$$

where N_O is Avogadro's number and n is the refractive index of the medium.

It should be noted that η_Y is the maximum thermodynamic efficiency obtained under reversible conditions, i.e., such that the rate of any photochemical reaction from D* is infinitesimally slow. Although η_Y has some theoretical interest, it has no practical interest since we are interested in maximizing the rate of a photochemical reaction from D* which will lead to the production of useful work. The rate of energy conversion by such a process can be defined as

$$P = J_e \cdot (1 - \phi_{loss}) \cdot \mu \tag{10}$$

where ϕ_{loss} represents the sum total quantum yield for all pro-

cesses which do not lead to conversion to work (e.g., fluorescence, non-radiative decay to D, etc.). Energy conversion from D*
will reduce the steady-state population of D* and hence μ will be
less than μ_{max}. P must be solved for a maximum such that ϕ_{loss} is
minimized while keeping μ as high as possible.

Using expressions derived by Ross and Hsiao (13) along with
their estimate that $\phi_{loss} \sim RT/\mu_{max}$, one can obtain the following
equation for the optimal rate of energy conversion

$$P = J_e \cdot [\mu_{max} - RT \ln (\frac{\mu_{max}}{RT})] \cdot [1 - \frac{RT}{\mu_{max}}] \qquad (11)$$

where it is assumed that non-radiative losses are insignificant
(i.e., $\phi_{internal\ conversion} = 0$). Then the maximum fraction of
the solar power which can be converted to useful work η_p is

$$\eta_p = \frac{P}{\int_0^\infty E(\lambda) d\lambda} \qquad (12)$$

η_p was calculated for AM 1.2 solar radiation and is plotted as
curve P in Fig. 2. It maximizes at 32% and 840 nm. It is interesting that ϕ_{loss} varies from 0.01 at 400 nm to 0.04 at 1500 nm
and is only 0.022 at the maximum of curve P; hence, most of the
difference between curves Y and P is due to a drop in μ from μ_{max}.

Curve P is quite general and η_p represents a realistic
maximum conversion efficiency for any device seeking to convert
solar energy directly to electricity or stored chemical energy.
Ross and Hsiao (13) have calculated curves E, Y and P for AM 0
solar radiation.

In general η_p increases with light intensity and hence better
performance could be obtained by concentrating sunlight (e.g., at
100 suns η_p increases from 32% to 36% at 840 nm); however, one
then has to contend with heat dissipation problems so the gain may
not be too significant since η_p drops rapidly with increasing
temperature (e.g., at 500 K η_p drops to 24% from 32% at 300 K at
840 nm).

Another way in which η_p may be increased is to use two separate photochemical sensitizers in two distinct photosystems, each
with a different range of spectral sensitivity. Analysis of AM
1.2 radiation (5) shows that a device in which one sensitizer absorbs all light with $\lambda \leq \lambda_1$ and a second sensitizer absorbs all
light in the range $\lambda_1 < \lambda \leq \lambda_2$ could achieve an efficiency of 44%
for $\lambda_1 = 830$ nm and $\lambda_2 = 1320$ nm. There is a wide range of
values of λ_1 and λ_2 for which η_p is above 40%.

For conversion of sunlight to electricity, there is no reason
why η_p cannot be approached. For example, a GaAs solar cell has
been reported (14) with an efficiency (AM 1.4) of 23% ($\lambda_g = 920$
nm). Indeed, an analysis of the maximum efficiency of a solar

cell based on a semiconductor model (15) reaches a conclusion very
similar to that considered here.

D.   A Kinetic Limitation on the Conversion of Light to Stored Chemical Energy

When we wish to convert and store solar energy as, for
example, in an endergonic photochemical reaction, an additional
kinetic requirement is imposed which can be seen by reference to
Fig. 1.   The conversion of R* to P must be an exergonic reaction
so that an activation energy E_r^* for the back reaction will be established (16).   Otherwise, P would have no stability for storage
or subsequent reactions leading to chemical storage.

If the photochemical step is first order (i.e., either unimolecular or bimolecular where the two reactant molecules are
restricted to react only with each other as in a solid state or on
a membrane), then we can apply unimolecular rate theory (17) to
estimate E_r

$$E_r^* \cong \frac{-kT}{e} \ln \left[\frac{h}{kT\tau}\right] \qquad \text{(in eV)} \qquad (13)$$

where τ is the lifetime of the primary products assuming that the
only reaction possible for the products is the back reaction.   For
example, if $\tau = 1$ ms, $E_r^* \cong 0.6$ eV, if $\tau = 1$ s, $E_r^* \cong 0.8$ eV; and if
$\tau = 10^3$ s, $E_r^* \cong 1.0$ eV.

The creation of this activation barrier must be done at the
expense of the excitation energy; the loss is constant and independent of λ_g.   Thus, we can define a rate of chemical yield C
(which, of course, must be less than P) as

$$C = E \cdot \eta_{chem} \qquad (14)$$

where E is defined by Eq. (2) and

$$\eta_{chem} = \frac{\Delta G}{E_g} \qquad (15)$$

where ΔG is the net free energy storage in the photochemical step
and E_g is defined in Fig. 1.   Then the maximum efficiency for the
rate of chemical storage will be

$$\eta_C = \frac{C}{\int_0^\infty E(\lambda)\,d\lambda} = \eta_E \cdot \eta_{chem} \qquad (16)$$

η_C is plotted as curve C in Fig. 2 assuming that $E_{R^*} - E_P = 0.6$
eV, 0.8 eV and 1.0 eV.   The curves maximize at 27% for $\lambda_g = 840$
nm, 21% for $\lambda_g = 750$ nm and 16% for $\lambda_g = 710$ nm, for 0.6 eV, 0.8

eV and 1.0 eV loss, respectively.

Again, the coupling of two photosystems in series can improve the yield. For example, with 0.8 eV loss from each photon, η_C is 29% for $\lambda_1 = 600$ nm and $\lambda_2 = 850$ nm.

One may argue that the loss could be smaller if τ were made shorter; however, secondary reactions necessary to stabilize the reaction products must be exergonic which will thus incur additional losses.

If the photochemical step is second order, as for the bimolecular reaction of two components in solution, then the loss can be quite small because of the short lifetime of the collision complex. However, in this case the back reaction will be virtually diffusion controlled and no net storage will occur. Anything which will slow down the back reaction must either involve exergonic secondary reactions or conversion of the process to first order (e.g. by confining the reactants to a surface or a membrane). Hence, *it is unlikely that one can avoid a loss of perhaps 0.6 - 1.0 eV in any photochemical endergonic step leading to chemical energy storage.*

It is interesting that in photosynthesis the energy loss in the primary photochemical step is ~ 0.8 eV for photosystems I and II of green plants and algae and also for photosynthetic bacteria (18). Also $\lambda_g \cong 700$ nm for green plant and algal photosynthesis, a value near the optimum for the 0.8 eV C curve in Fig. 2.

E. An Estimate of Solar Energy Conversion Efficiency

Curve P in Fig. 2 represents the ideal thermodynamic limit for conversion efficiency and curve C sets an approximate limit for the efficiency of conversion to stored chemical energy. There are, however, other loss factors to be considered which vary according to the device. These loss factors have been considered in considerable detail for photovoltaic devices (15, 19) and estimates for the ultimate achievable conversion efficiency to electricity vary from 25 - 28%; however, only recently have efficiencies been considered for conversion to stored chemical energy (5). In this case

$$\eta_{storage} = \eta_{abs}\eta_E\eta_{chem}\eta_\phi\eta_{coll} \tag{17}$$

where η_{abs} is the fraction of incident photons with $\lambda \leq \lambda_g$ which are absorbed; η_E and η_{chem} are defined by Eqs. (3) and (15) respectively; η_ϕ is the quantum yield for the photochemical step(s); η_{coll} is the fraction of product produced which can be collected and stored.

η_{abs} is unlikely to be greater than ~ 0.75 for most absorbers and η_ϕ and η_{coll} are unlikely to be greater than 0.9 each. Thus, if we take all of the factors together with the maximum value of $\eta_E\eta_{chem} = 0.21$ from Fig. 2, then we find that the net yield of product in a photochemical energy storage reaction is unlikely to be greater than 12 - 13%. This figure should not be discouraging because photosynthesis, which must be considered a very useful process, has a net efficiency of $\sim$ 6% under ideal conditions (18) and 1 - 3% under actual field conditions (20).

A potentially more serious problem than that of low efficiency is the requirement that the photochemical absorber operate without any significant side reactions. For example, if the quantum yield for side reactions were 1%, then after only 100 cycles the concentration of the absorber would have decreased to $\sim$ 37% of its original concentration. It is interesting that in photosynthesis each chlorophyll molecule processes at least 10^5 photons in its lifetime in a leaf. This means that the quantum yield for reactions leading to the degradation of chlorophyll must be less than 10^{-5}!

F. Threshold Wavelength

For photochemical reactions leading to chemical energy storage, there will be a maximum wavelength (λ_{max}), called the threshold wavelength, which will be capable of initiating the endergonic photochemical reaction. λ_{max} can be calculated from (16)

$$\lambda_{max} = \frac{nN_o hc\eta_{chem}}{\Delta G} \qquad (18)$$

where n is the number of photons which must be absorbed to carry out the overall reaction and ΔG is the free energy storage per mole of product formed. η_{chem} will be given by

$$\eta_{chem} = \frac{\Delta G}{\Delta G + E_r^* neN_o} \qquad (19)$$

where E_r^* is the activation energy for the back reaction in eV.

G. Some Potentially Useful Reactions

Photochemical reactions leading to the conversion and storage of solar energy can be divided into five types.

1. Molecular Energy Storage Reactions. This type of photochemical reaction involves formation of new bonds either intramolecularly (to form an isomer of the reactant molecule, e.g., a cis-trans isomerization about a double bond) or bimolecularly to form an addition compound (e.g., the dimerization of anthracene). The reactions studied so far almost all involve unsaturated organic molecules. The endergonic or energy-storage feature of these reactions often arises from excessive "strain" induced in the product molecule or loss of resonance energy. The back reaction to reform the reactant(s) in most cases yields only heat and thus this type of reaction should be considered as a possible means for the latent storage of heat. Sasse (21) has written a comprehensive review of this field. Unfortunately, most of the reactions studied involve ultraviolet absorbers and hence the overall storage efficiency of solar energy would be less than 1 - 2%. Nevertheless, it may be possible to extend the absorption

limit well into the visible by molecular modifications.

Economic analyses have been carried out on a hybrid thermal and photochemical collector and storage system by Talbert et al. (22). Conversion efficiency, energy storage capacity and life-cycle costs were the primary bases of comparison. To a certain extent, chemical efficiency can be traded off against energy-storage capability. However, there are limits and it appears that a specific heat storage capacity of 300 - 400 Jg^{-1} with an overall chemical storage efficiency of $\sim$ 20% will be required. It may be possible to meet the former criterion but reference to curve C of Fig. 2 shows that the latter criterion will be very difficult to meet.

   2.   Homolytic Bond Fission Reactions. The homolytic fission of a chemical bond

$$AB \rightarrow A\cdot + B\cdot \tag{19}$$

represents a particularly simple photochemical reaction and is always endergonic. Unfortunately, as Moggi points out (23) the products of the primary step are free radicals which are usually very reactive and undergo further reactions which degrade part of the energy available in the first step. Another difficulty is that the absorbed photon must have an energy greater than the AB bond energy. Thus, if we wish the molecule to undergo photochem-ical homolytic fission with solar wavelengths even barely into the visible, then the AB bond energy must be less than $\sim$ 300 kJ mol^{-1}. This rules out most chemical bonds.

   Nevertheless, there are a few reactions which satisfy the stringent criteria considered above, two of which have been studied as possible energy storage reactions. One of these is the photolysis of nitrosyl chloride (24)

$$NOCl \rightarrow NO + 1/2\ Cl_2$$

This reaction has the advantage that it is sensitized by visible light out to 635 nm and the quantum yield is high. Unfortunately, it is difficult to prevent the back reaction and $\Delta G°$ for the reaction is only 20.5 kJ mol^{-1}. Nevertheless, it is theoretically possible to store up to $\sim$ 4% of the input solar energy so more work on this reaction would seem to be warranted.

   The other interesting reaction is the decomposition of ferric bromide ions

$$FeBr^{2+} \rightarrow Fe^{2+} + 1/2\ Br_2 \tag{21}$$

This reaction has been studied by Chen et al. (25) and has the advantage that the Br_2 can be easily removed in a stream of nitro-gen; however, the quantum yield is only 1 - 2% and $\Delta G°$ is only 28.5 kJ mol^{-1}. Hence, only a very small fraction ($\sim$ 0.05%) of solar energy can be stored.

It is clear that if homolytic bond fission reactions are going to be useful, they must have almost perfect quantum yields and a $\Delta G°$ between 100 and 150 kJ mol^{-1}. One reaction which might work is

$$NOF \rightarrow NO + 1/2 \ F_2 \qquad (22)$$

$\Delta G° = 137.6$ kJ mol^{-1}. Theoretically, this reaction could have an efficiency of over 8%. However, the corrosive nature of fluorine gas would be a problem.

3. <u>Homogeneous Redox Reactions Leading to the Generation of Solar Fuels</u>. The two previous reaction types considered involved mostly unimolecular reactions although some bimolecular addition reactions were included in molecular energy storage reactions. Unimolecular reactions have the advantage of only one reactant, but, as we have seen, there are other negative features which make them unattractive. One of the major difficulties is that, with a few exceptions, the whole energy-storage process must be accomplished with one photon and only a small fraction of the photon energy ends up as stored chemical energy. Because of the fact that solar energy contains mostly relatively low energy photons, any process which can integrate many photons to drive an endergonic chemical reaction will have an inherent advantage. For example, photosynthesis achieves its reaction by integrating at least eight photons for every molecule of CO_2 incorporated into carbohydrates.

Taking our cue again from photosynthesis, the primary photochemical reactions should be simple one-electron transfer reactions producing an oxidant and a reductant. That is

$$A + B \xrightarrow{h\nu} A* + B \rightarrow A^+ + B^- \qquad (23a)$$

or

$$A + B \xrightarrow{h\nu} A + B* \rightarrow A^+ + B^- \qquad (23b)$$

where A* and B* are the reactive excited states of A and B, respectively.

As noted in Section D, ~ 0.8 eV must be lost in converting the energy of the excited state into stable products. An additional requirement in this case is the need for a mechanism to store electrochemical equivalents as most useful redox reactions require the transfer of more than one electron. For example, in photosynthesis, the water-splitting enzyme accumulates four positive charges necessary to oxidize two water molecules to one molecule of oxygen while the ferredoxin-NADP-reductase enzyme carries out the function of accumulating two negative charges necessary to produce one molecule of NADPH which provides the reducing power to reduce CO_2 to carbohydrates. These charge-storage catalysts are very important because they provide a concerted pathway to the ultimate products without requiring that high energy and very reactive free radical intermediates be formed.

Although there are many possible redox reactions, which, in principle, could be used to store chemical energy, it would be most desirable from a practical standpoint to generate a fuel which is already in use from fossil fuel sources. Thus for the purposes of this article, I will adopt a limited definition of a fuel as *any reduced chemical substance produced as a result of an endergonic photochemical reaction, which on reaction with oxygen will release the stored chemical energy*. A general reaction producing a fuel F is hence written as

$$A + B \xrightarrow{h\nu} F + O_2 \qquad (24)$$

where the stoichiometry need not be that given in Eq. (24) and the back reaction serves to release the stored energy.

Taking into account the requirements listed in Section A, it would be desirable if the reactants A and B be compounds which are very cheap and readily available. Naturally, the constituents of the atmosphere and liquid water fill this requirement admirably. Table 1 lists most of the endergonic fuel generation reactions which involve N_2, CO_2 and H_2O as reactants including the reaction of photosynthesis. It is significant that the potential difference ΔE°, which is the potential stored per electron transferred, is between 1.02 V and 1.48 V for all of the reactions in Table 1. Thus, the energy requirements for the photochemistry are about the same for each of these reactions. We immediately see that the reaction of photosynthesis (reaction 9 of Table 1) is in trouble for one photosystem because λ_{max} is known to be 700 nm. The implication is that the photosynthesis reaction cannot be operated at 700 nm with one photochemical system (i.e., one electron transferred per photon absorbed) (18). Indeed, in order to be able to use the longer wavelength photons, photosynthesis has employed two photosystems operating in series so that eight photons are used to drive the reaction instead of four. With two photosystems λ_{max} is 872 nm (column 6 of Table 1) which leaves plenty of scope for an absorber at 700 nm.

Although λ_{max} in column 5 of Table 1 represents the maximum threshold for the photochemistry where only one photosystem is employed, the true λ_{max} is almost certain to be considerably less than the values in Table 1 because of subsequent secondary reactions which must be exergonic and hence will lose more energy. Thus, we must conclude that it is likely that more success will be achieved if attention is given to photochemical processes employing two coupled reactions so that two photons are used for every electron transferred in the ultimate reaction (16). λ_{max} values for two photosystems are given in column 6 of Table 1.

Of all the photochemical energy storage reactions proposed,

Table I. Some Endergonic Fuel Generation Reactions Starting with N_2, CO_2 and H_2O

Reaction	$\Delta H°$ (kJ mol⁻¹) [a]	$\Delta G°$ (kJ mol⁻¹) [a]	n [b]	$\Delta E°$ (V)	λ_{max} (nm) [c] one photosystem	λ_{max} (nm) [c] two photosystems
1. $H_2O(\ell) \rightarrow H_2(g) + \frac{1}{2} O_2(g)$	286	237	2	1.23	611	877
2. $CO_2(g) \rightarrow C(s) + O_2(g)$	394	394	4	1.02	681	946
3. $CO_2(g) \rightarrow CO(g) + \frac{1}{2} O_2(g)$	283	257	2	1.33	581	845
4. $CO_2(g) + H_2O(\ell) \rightarrow HCOOH(\ell) + \frac{1}{2} O_2(g)$	270	286	2	1.48	543	804
5. $CO_2(g) + H_2O(\ell) \rightarrow HCHO(g) + O_2(g)$	563	522	4	1.35	576	840
6. $CO_2(g) + 2 H_2O(\ell) \rightarrow CH_3OH(\ell) + \frac{3}{2} O_2(g)$	727	703	6	1.21	616	881
7. $CO_2(g) + 2 H_2O(\ell) \rightarrow CH_4(g) + 2 O_2(g)$	890	818	8	1.06	667	932
8. $N_2(g) + 3 H_2O(\ell) \rightarrow 2 NH_3(g) + \frac{3}{2} O_2(g)$	765	678	6	1.17	629	895
9. $CO_2(g) + H_2O(\ell) \rightarrow \frac{1}{6} C_6H_{12}O_6(s) + O_2(g)$	467	480	4	1.24	607	872

a. All thermodynamic data have been obtained from "Selected Values of Chemical Thermodynamic Properties" Part 1 National Bureau of Standards Cricular 500 U.S. Government Printing Office, Washington, D.C. (1961) except data for C6H12O6(s) which were obtained from D.R. Stull, E.F. Westrum and G.C. Sinke, "The Chemical Thermodynamics of Organic Compounds", John Wiley and Sons, New York, 1969, p. 680.

b. n is the number of electrons which should be transferred in an electrochemical reaction for the reaction as written.

c. λ_{max} is calculated from Eq. (18) assuming that η_{chem} is given by Eq. (19) and n is doubled for two photosystems.

the production of hydrogen and oxygen from water (reaction 1 in Table 1) is one of the most attractive. Hydrogen is an almost ideal fuel and the starting material is certainly cheap. Several authors have considered the possibilities for a homogeneous photochemical water-splitting reaction (1-5, 16, 23, 26-30). For the reasons noted above, it is unlikely that this reaction can be sensitized except with blue and ultraviolet light where only one photosystem is employed. Bolton (5, 16) has proposed a scheme whereby the reaction might be carried out using two photosystems. Nevertheless, it may be that in the long run a photochemical method to convert CO_2 to a useful fuel will prove more valuable. After all, hydrogen can be obtained by electrolysis of water using photovoltaic or photoelectrochemical cells to provide the electricity; whereas, there is no known method of producing carbonacious fuels by electrolysis. One can imagine a photochemical plant using solar energy and stack gas from a fossil fuel power plant to generate some of the fuels listed in Table 1.

   4. <u>Photoelectrochemical Generation of Solar Fuels</u>. In some cases it is possible to employ a photosensitive electrode, usually a semiconductor electrode, to carry out essentially an electrolysis reaction. The first successful experiment was described by Rujishima and Honda (31). This process is called photoelectrolysis (when no external electrical source is required) and photoassisted electrolysis (when a small external potential is required for the reaction to go). The only reaction studied so far is that of the decomposition of water into hydrogen and oxygen. Since this is a chemical energy storage process, the same limitations and considerations which were developed for homogeneous redox reactions in the previous section apply also to photoelectrolysis processes. The analog to a single photochemical system is a cell in which only one electrode is illuminated. Indeed, the only cells which work without the need for an external electrical potential, are those which require ultraviolet illumination. The electrochemical analog to two photochemical system is a cell in which both electrodes participate in the photochemistry. The requirement for two photosensitive electrodes was pointed out by Manassen et al. (32). Nozik (33) has proposed such a cell and has had some limited success. Recently, Fong et al. (34) described a photogalvanic water splitting system using chlorophyll a dihydrate where λ_{max} is 740 nm. From reaction 1 of Table 1, we see that this λ_{max} is well beyond the limit for a single photosystem. It may be that this electrode is operating by a biphotonic mechanism perhaps utilizing the triplet state of chlorophyll as the intermediate state. The field of photoelectrolysis has been reviewed recently by Gerischer (35).

   5. <u>Photochemical Systems Designed to Generate Electricity</u>. There are three types of conversion devices which involve the production of electricity driven by a photochemical reaction.

a. Solid-state photovoltaic cells: Semiconductor solar
cells such as the silicon, gallium arsenide and cadmium sulfide
cells are the most developed and commercially viable photochemical devices available today. Efficiencies are high and the only
impediment to large-scale commercialization is the cost which is
currently about $10/peak watt (36). Recently, some interesting
work has been reported on organic solar cells which involve an
organic dye, such as chlorophyll, to initiate the photocurrent
(37). Efficiencies are still low but a cell using monomolecular
layers which we have developed (38) shows some promise as it is
possible to use "molecular engineering" to improve the efficiency.

b. Photogalvanic cells: These are defined as cells in which
a dye in solution undergoes a photochemical reaction to produce
products which then migrate to electrodes where the conversion to
electricity occurs. Since a chemical intermediate must be formed,
the kinetic limit (curves C of Fig. 2) apply. This field has been
weel reviewed by Lichtin (39). A theoretical analysis by Albery
and Archer (40) indicates that it may be very difficult to achieve
very high efficiencies with photogalvanic cells.

c. Regenerative photoelectrochemical cells: These cells are
very much like the photoelectrolysis cells considered above except
that no net chemistry occurs since products produced at one electrode are consumed at the other. These cells are sometimes called
"wet" solar cells. The photochemistry occurs in photosensitive
semiconductor electrodes. Gerisher (35) has reviewed the principles and applications of these cells.

I. Conclusions

Within the kinetic and thermodynamic limitations on the
conversion of light energy to chemical energy, I have shown that
a reasonable efficiency goal would be $\sim$ 25 - 28% for conversion to
electricity and $\sim$ 10 - 13% for storage as chemical energy. Five
types of photochemical converters have been defined and described
with examples where possible.
    This field is in its infancy - clearly, much more basic and
mission-oriented research will be necessary to establish if
workable and economic fuel-generation systems and electrical
generation systems can be developed. The challenge now is to
chemists, physicists and biologists to develope systems that at
least work in the laboratory. Only then can meaningful economic
analyses be made. Hopefully this article will help to provide
some guidelines and objectives for the research that must be done.

Literature Cited

1. Levine, S., Halter, H. and Mannis, F., Solar Energy (1958), 2, pp. 11.
2. Calvert, J.G., "Photochemical Processes for Utilization of Solar Energy" in Introduction to the Utilization of Solar Energy, ed. by A.M. Zarem and D.D. Erway, pp. 190-210, McGraw-Hill, New York, 1964.
3. Balzani, V., Moggi, L., Manfrin, M.F., Bolletta, F. and Gleria, M., Science (1975), 189, pp. 852-856.
4. Porter, G. and Archer, M.D., Interdiscip. Sci. Rev. (1976), 1, pp. 119-143.
5. Bolton, J.R., Science (in press).
6. Boer, K.W., Solar Energy (1977), 19, pp. 525-538.
7. Ross, R.T., J. Chem. Phys. (1966), 45, pp. 1-7; ibid (1967), 46, pp. 4590-4593.
8. Ross, R.T. and Calvin, M., Biophys. J. (1967), 7, pp. 595-614.
9. Ross, R.T., Anderson, R.J. and Hsiao, T.-L., Photochem. Photobiol. (1976), 24, pp. 267-278.
10. Duysens, L.N.M., in "Photochemical Apparatus", Brookhaven Symposium in Biology No. 11, pp. 18, 1958.
11. Mortimer, R.G. and Mazo, R.M., J. Chem. Phys. (1961), 35, pp. 1013-1018.
12. Almgren, M., Photochem. Photobiol. (in press).
13. Ross, R.T. and Hsiao, T.-L., J. Appl. Phys. (1977), 48, pp. 4783-4785.
14. James, L.W. and Moon, R.L., Appl. Phys. Lett. (1975), 26, pp. 467-470.
15. Shockley, W. and Queisser, H.J., J. Appl. Phys. (1961), 32, pp. 510-519.
16. Bolton, J.R., Solar Energy (1978), 20, pp. 181-183; J. Solid State Chem. (1977), 22, pp. 3-8.
17. Laidler, K.J., "Chemical Kinetics" 2nd Ed., pp. 165, McGraw-Hill, New York, 1965.
18. Bolton, J.R., in "Photosynthesis 77: Proceedings of the 4th International Congress on Photosynthesis", D.O. Hall, J. Coombs and T.W. Goodwin, eds., pp. 621-634, The Biomedical Society, London, 1978.
19. Loferski, J.J., J. Appl. Phys. (1956), 27, pp. 777-784; Wysocki, J.J. and Rappaport, P., J. Appl. Phys. (1960), 31, pp. 571-578; Wolf, M., Energy Conv. (1971), 11, pp. 63-73.
20. Hall, D.O., in "Solar Power and Fuels", Proceedings of the First International Conference on the Photochemical Conversion and Storage of Solar Energy, London, Canada, J.R. Bolton ed., pp. 27-52, Academic Press, New York, 1977.
21. Sasse, W.H.F., in "Solar Power and Fuels", Proceedings of the First International Conference on the Photochemical Conversion and Storage of Solar Energy, London, Canada, J.R. Bolton ed., pp. 227-248, Academic Press, New York, 1977.

22. Talbert, S.G., Frieling, D.H., Eibling, J.A. and Nathan, R.W., Chem. Tech. (1976), 6, pp. 118-122; Solar Energy (1975), 17, pp. 367-372.
23. Moggi, L., in "Solar Power and Fuels", Proceedings of the First International Conference on the Photochemical Conversion and Storage of Solar Energy, London, Canada, J.R. Bolton, ed., pp. 147-165, Academic Press, New York, 1977.
24. Kistiakowsky, G.B., J. Amer. Chem. Soc. (1930), 52, pp. 102-108; Neuwirth, O.S., J. Phys. Chem. (1959), 63, pp. 17-19.
25. Chen, S.-N., Lichtin, N.N. and Stein, G., Science (1975), 190, pp. 879-880.
26. Heidt, L.J. and McMillan, A.F., Science (1953), 117, pp. 75-76.
27. Marcus, R.J., Science (1956), 123, pp. 399-405.
28. Paleocrassas, S.N., Solar Energy (1974), 16, pp. 45-51.
29. Heidt, L.J., in "Solar Energy Research", F. Daniels and A.D. Duffie, eds., pp. 203-220, Univ. Wisc. Press, Madison, Wisc., 1953.
30. Stein, G., Israel J. Chem. (1975), 14, pp. 213-225.
31. Fujishima, A. and Honda, K., Bull. Chem. Soc. Japan (1971), 44, pp. 1148-1150; Nature (1972), 238, pp. 37-38.
32. Manassen, J., Cahen, D., Hodes, G. and Sofer, A., Nature (1976), 263, pp. 97-100.
33. Nozik, A.J., Appl. Phys. Lett. (1976), 29, pp. 150-153.
34. Fong, F.K., Polles, J.S., Galloway, L. and Fruge, D.R., J. Amer. Chem. Soc. (1977), 99, pp. 5802-5804.
35. Gerischer, H., in "Solar Power and Fuels", Proceedings of the First International Conference on the Photochemical Conversion and Storage of Solar Energy, London, Canada, J.R. Bolton, ed., pp. 77-117, Academic Press, New York, 1977.
36. For reviews of photovoltaic principles and applications, consult Merrigan, J.A., "Sunlight to Electricity - Prospects for Solar Energy Conversion by Photovoltaics", MIT Press, Cambridge, Mass., 1975; "Solar Cells", C.E. Backus, ed., IEEE Press, New York, 1976; Hovel, H.J., "Solar Cells for Terrestrial Applications", Proceedings of the Conference "Sharing the Sun", Winnipeg, Canada, Vol. 6, pp. 1, Pergamon Press, London, 1976; Hovel, H.J., "Solar Cells", Vol. 11 of Semiconductors and Semimetals, Beer and Willardson, eds., Academic Press, New York, 1976.
37. Tang, C.W. and Albrecht, A.C., J. Chem. Phys. (1975), 63, pp. 953.
38. Janzen, A.F. and Bolton, J.R., unpublished results.
39. Lichtin, N.N., in "Solar Power and Fuels", Proceedings of the First International Conference on the Conversion and Storage of Solar Energy, London, Canada, J.R. Bolton, ed., pp. 119-142, Academic Press, New York, 1977.
40. Albery, W.J. and Archer, M.D., J. Electrochem. Soc. (1977), 124, pp. 688.

RECEIVED September 25, 1978.

Photoelectrolysis of Aqueous Solutions to Hydrogen—An Approach to Solar Energy Storage

FRANK R. SMITH

Chemistry Department, Memorial University of Newfoundland,
St. John's, Newfoundland, Canada A1B 3X7

It is generally agreed that electrolysis of aqueous solutions offers the best prospect for the production of hydrogen from water, because of the easy separation of the H_2 and O_2 products and because of the relatively low energy consumption if catalytically active metal electrodes are used. Thus, the minimum energy requirements are those for which water, hydrogen and oxygen, each at 1 atmosphere pressure, are in equilibrium:

$$H_2O(l) = H_2(g) + \tfrac{1}{2} O_2(g) \qquad E_{reversible} = 1.23 \text{ V at 298 K}$$

This equilibrium e.m.f. increases with hydrogen and oxygen partial pressures:

$$E_{rev} = 1.23 \text{ V} + \frac{2.303 \text{ RT}}{F} \; \log_{10} P_{H_2} \, P_{O_2}^{\frac{1}{2}} \text{ at 298 K.} \tag{1}$$

The energy requirement of 1.23 eV per half-molecule of H_2 translates to a minimum of 10.6 MJ per cubic metre at S.T.P. of hydrogen, whereas practical electrolysers operate somewhere between 14 and 25 MJ m^{-3} (1). This is because of the need to carry out electrolysis at a finite rate, that is to say at potential differences greater than the equilibrium conditions referred to above. The voltages are wasted in two principal ways: overvoltages associated with hydrogen and oxygen evolution, η_c and η_a, respectively, and resistive losses. Since the exchange current densities for oxygen evolution, i'_o, are in general smaller than those for hydrogen evolution, i_o, the overvoltages associated with the former are larger than those of the latter. Both increase in a logarithmic fashion with current density, i:

$$\eta_c = \eta_{H_2} = b \log_{10}(i/i_o) \qquad (b \sim 30 \text{ mV at 300K}) \tag{2}$$

$$\eta_a = \eta_{O_2} = b' \log_{10}(i/i'_o) \qquad (b' \sim 60 \text{ mV at 300K}) \tag{3}$$

The resistive losses (the iR drops in the solution between the
electrodes) are more significant because they increase linearly
with current density. Bringing the electrodes closer together
should diminish these losses, in principle, but electrolysis at
high pressure is even more advantageous in decreasing resistive
losses (e.g. at 100 atmospheres by 0.4 V) because the gas bubbles
are smaller, this decrease more than offsetting the 0.1 V increase
in E_{rev} (2). Iron or nickel-plated steel electrodes are usually
used, these metals being quite good catalysts, even if they are
inferior to the much more expensive platinum. Recently, the use
of a thin solid ion-exchange resin electrolyte and finely divided
platinum cathode has been advocated for lowering the energy
consumption in practical electrolytic hydrogen generation (3).

Photo-effects at Metals and Semiconductors

        In order to diminish the energy which has to be supplied from
an electrical power source to liberate hydrogen from water, one
may look for assistance in the form of light energy. Photo-
effects have been observed at metal-aqueous solution interfaces,
e.g. at mercury, but these involve production of hydrated
electrons and their subsequent reactions with electron scavengers
(4). They are not likely to be of interest in the present
context. Instead, one must consider semiconductor-aqueous
solution interfaces, where photo-currents are known to occur,
enabling electrochemical reactions to occur in illumination which
hardly occur at all in the dark. Thus, hydrogen evolution at
p-type Ge (5,6) and at p-type Se (7) occurs with increased current
density (rate) or with decreased overvoltage upon illumination.
In the case of selenium, green light diminished the overvoltage
much more than did red light, only the former being of energy $\geq$
the band gap of the semiconductor (1.8 eV (8)). Whereas the
reduction of H^+ was shown to require transfer of conduction band
electrons from n-type Ge in the dark or of photo-generated
electrons (from the conduction band) of illuminated p-type Ge, the
reduction of O_2 required hole transfer to the valence band (9).
Although oxygen liberation was not demonstrated using germanium
electrodes, because of the intervention of anodic dissolution of
the semiconductor, a process involving hole transfer from the
valence band, it is evident that the reverse reaction to O_2
reduction also involves hole transfer, that is to say it would
occur at a p-type semiconductor like germanium in the dark, or at
the same n-type semiconductor when illuminated.
        In fact, in 1972, Fujishima and Honda (10) demonstrated that
O_2 evolution on n-type TiO_2 occurs as a photocurrent, proportional
to the light intensity (Figure 1) of wavelengths less than 415 nm,
i.e. for photon energies equal to or greater than the band gap of
TiO_2 : 3.0 eV. In this work and that of Ohnishi et al. (11) a
platinum black metal cathode was connected in an external circuit
to an indium contact on the back side of the photo-anode (see

Figure 2). Hydrogen was evolved at the cathode, the over-all process being the decomposition of water. This process, involving an applied potential difference between the anode and cathode, we shall refer to as photo-assisted electrolysis.

Energy Conditions for Photo-assisted Electrolysis

Referring to Figure 3, evidence exists for placement of the Fermi levels (chemical potentials) of the redox reactions involving H_2, H_2O and O_2 roughly at the positions shown relative to the energies of the conduction band minimum and valence band maximum of the semiconductor, E_c and E_v, respectively. This picture takes the electron in a vacuum at infinity as the zero of energy. On this basis, the Fermi level for the reaction

$$H^+_{aq} + e^-_{vac} = \tfrac{1}{2} H_2(g)$$

has the value $E^o_{F(H_2O/H_2)} \sim -4.5$ eV for unit activity of all the species ($\underline{12}$). This E^o_F is the standard chemical potential per electron transferred, i.e. $\Delta G^o / n N_o$, for the reaction, where ΔG^o is the standard Gibbs free energy of the reaction, n the number of electrons transferred and N_o Avogadro's number. On the same basis, the reduction of oxygen under standard conditions

$$\tfrac{1}{4} O_2(g) + \tfrac{1}{2} H_2O(l) + e^-_{vac} = OH^-_{aq}$$

has a Fermi level, $E^o_{F(O_2/H_2O)} \sim -5.73$ eV, or 1.23 eV below that for the hydrogen ion. Changes in concentrations of reactants and products change the Fermi levels for redox reactions in accordance with a modified form of Nernst equation, e.g. for the hydrogen reaction,

$$E_{F(H_2O/H_2)} = E^o_{F(H_2O/H_2)} + kT \ln \frac{a^{\tfrac{1}{2}}_{H_2}}{a_{H^+}} \qquad (4)$$

At equilibrium in the dark, the Fermi levels for electrons and holes in a particular sample of semiconductor are coincident, being near the top of the forbidden gap in an n-type and near the bottom of the gap in a p-type semiconductor. Figure 4 illustrates the region near the surface of an n-type semi-conductor, the Fermi levels for electrons and holes, coincident up to the electrode-solution interface, being denoted $_nE_F$ and $_pE_F$, respectively. In Figure 3 the Fermi levels for electrons and holes are separated as a result of the absorption of photons of energy, $h\nu$, greater than the semiconductor band gap. This absorption of energy leads to the generation of electrons and holes (one of each for each photon absorbed), in the volume of the semiconductor penetrated by the light. The degree of

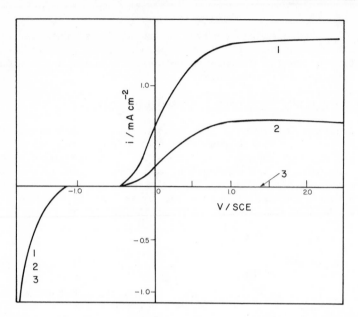

Nature

Figure 1. Photocurrent/electrode potential curves for n-type TiO₂ single crystal with ohmic-indium contact on back side. Potentials measured with respect to saturated KCl calomel electrode with a platinum black counter electrode. Light intensity increasing in order 3,2,1. Wavelength, 415 nm or less. Exposed surface of TiO₂ crystal: (001) (10).

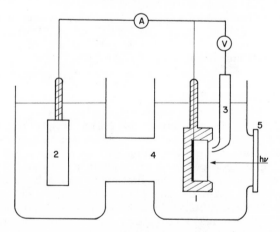

Verlag Chemie GmbH

Figure 2. Cell and circuit used in experiments like those in Figure 1. (1)Illuminated TiO₂ electrode; (2) platinum counter electrode in the dark; (3) reference electrode (SCE); (4) buffered electrolyte solution; (5) quartz window for UV light; (A) ammeter; (V) voltmeter (11).

separation of the quasi-Fermi levels, E_{nF}^* and E_{pF}^* increases with the light intensity, as Figure 5 illustrates.

It might be thought that photogeneration of electrons and holes in a semiconductor with normally a small population of one or other (or both) charge carrier would result in both hydrogen and oxygen production at the interface, through the reactions:

$$H_{aq}^+ + e^- \text{ (conduction band)} = \tfrac{1}{2} H_2(g)$$

$$OH_{aq}^- + h^+ \text{ (valence band)} = \tfrac{1}{4} O_2(g) + \tfrac{1}{2} H_2O(l)$$

Gerischer (13) has termed such a process "photocatalytic action of a semiconductor electrode". The reason that such processes hardly occur is that they are interfacial reactions and, in travelling from the region beneath the surface to the interface, the electrons and holes have many encounters, these encounters leading to recombination and re-emission of light. Referring to Figure 3, necessary conditions for the reactions above are fulfilled when, respectively,
1) the Fermi level for electrons lies above that of the H2O/H2 redox system
and
ii) the Fermi level for holes lies below that of the O2/H2O system.

Unfortunately, these conditions are not sufficient. It is necessary to find some way to separate the charge carriers because their interaction is so strong and their recombination so rapid that they must not be permitted to occupy the same region of the crystal. This is not such a difficult condition to meet as might be at first thought. Figure 4 and Figure 5 are illustrative of one approach, that of using an n-type semi-conductor with a surface having a considerably decreased electron population, a so-called depletion layer. This material would form the anode for oxygen generation, hydrogen generation occurring at the metal cathode. These are the conditions already exemplified by the work of Fujishima and Honda (10) in Figures 1 and 2. Two alternative approaches are possible, use of a p-type semiconductor cathode and a metal anode or use of a p-type semiconductor cathode and an n-type anode in combination, in each case the semiconductor having a depletion layer at its surface. Each approach will be outlined, beginning with the n-type anode.

n-type Semiconductor in the Dark and Under Illumination

Figures 4 and 5 show an n-type semiconductor with a surface depletion layer, the energy bands bending upwards as the surface is approached from the interior. Under illumination (Figure 5) the band bending is diminished because of the photo-generation

226   CHEMISTRY FOR ENERGY

Figure 3. Schematic of a semiconductor–aqueous electrolyte solution interface, ignoring band bending. E_c and E_v are the band edges of the conduction and valence bands, respectively. $E_{F(H_2O/H_2)}$ and $E_{F(O_2/H_2O)}$ are the Fermi levels in the solution for the redox reactions indicated. The quasi-Fermi levels with illumination by light of energy hν are designated $_nE_F^*$ and $_pE_F^*$ respectively, for electrons and holes (13).

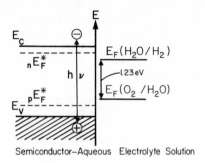

Semiconductor–Aqueous Electrolyte Solution

Figure 4. Semiconductor–electrolyte solution interface in the dark. An n-type semiconductor with a depletion layer at the surface is illustrated. E is electron energy, $_nE_F$ and $_pE_F$ are the equal Fermi levels for electrons and holes at equilibrium, other symbols as in Figure 3 (13).

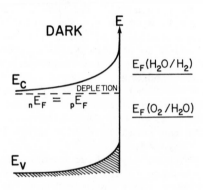

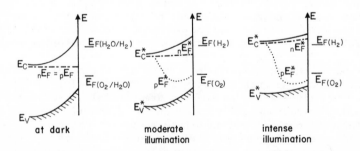

Figure 5. n-Type semiconductor–electrolyte solution interface with a surface depletion layer, in the dark and with two intensities of illumination. Symbols as in Figure 3 and 4 with E_c^* and E_v^* the band edges of the conduction and valence bands, respectively, under illumination, and $E_{F(H_2)}$ and $E_{F(O_2)}$ abbreviations for $E_{F(H_2O/H_2)}$ and $E_{F(O_2/H_2O)}$, respectively. The quasi-Fermi levels $_nE_F^*$ and $_pE_F^*$ are at different positions in the surface region than in the bulk as a result of the limited penetration of light into the interior. Fermi levels in solution as in Figures 3 and 4 (13).

of carriers in the region near the surface. The band gap remains the same: $E_c^* - E_v^* = E_c - E_v$. Whereas in the dark the concentrations of electrons, n, and of holes, p, are related to the (equal) numbers of electrons and holes in an intrinsic semi-conductor, n_i, by the mass action law:

$$n\,p = n_i^2 \tag{5}$$

under conditions of illumination the concentrations are increased to values n^* and p^* where $n^* - n$ and $p^* - p$ depend on the light intensity and the quasi-Fermi levels differ from the equilibrium (dark) values:

$$_nE_F^* - {}_nE_F = kT \ln \frac{n^*}{n} \tag{6}$$

$$_pE_F^* - {}_pE_F = kT \ln \frac{p^*}{p} \tag{7}$$

For an n-type semiconductor the quasi-Fermi level for electrons will not shift very much except at the surface which formerly was depleted of electrons, because elsewhere n^* will not be greatly in excess of n. For holes, however, the concentration p^* will be much larger than p where light penetrates into the semiconductor. For this reason, the quasi-Fermi level for holes $_pE_F^*$, departs markedly from its former value when light of sufficiently high frequency is incident upon the material.

Photo-assisted Electrolysis of Water Using n-type Anode and Metal Cathode

Four criteria have to be met for successful photo-assistance with the electrolysis of water to hydrogen and oxygen, assuming that the semiconductor band gap exceeds 1.23 eV.
First, there must be upward bending of the energy bands at the interface, <u>even under intense illumination.</u>
Second, the quasi-Fermi level for electrons must lie above that for the H_2O/H_2 redox system.
Third, the light intensity must be high enough to split sufficiently the quasi-Fermi levels for electrons and holes, i.e. by more than 1.23 eV, such that the level for holes at the interface is below that for the O_2/H_2O redox system, as shown in Figure 6.
Fourth, the favoured anodic reaction at the semiconductor must be O_2 evolution from water, rather than some anodic dissolution process in which the semiconductor breaks down, as happens with Ge (<u>9</u>), GaP (<u>14,15,16</u>) CdS or even ZnO (<u>13</u>).
In Figure 6, the case of n-type TiO_2 and a metal cathode is depicted, with an applied potential difference $(E_F - E_F')/e$, either to ensure that the Fermi level of electrons in the metal is higher than the H_2O/H_2 redox system so that hydrogen

evolution occurs at all, or to increase the rate of such
evolution (and that of O_2 as well). Holes, generated near the
semiconductor-electrolyte interface, travel towards this interface, some being lost by recombination while the remainder react:

$$OH_{aq}^- + h^+(TiO_2, \text{ valence}) = \tfrac{1}{4} O_2(g) + \tfrac{1}{2} H_2O(l)$$

Electrons, generated near the semiconductor-electrolyte interface
are unable to stay in this region because of the electric field
there which drives them into the bulk of the TiO_2 crystal, out
through the metallic contact, the external circuit (where the
photo-current may be measured) and into the catalytically active
metal. At the interface of this metal with the electrolyte
solution, reaction occurs:

$$H_{aq}^+ + e^-(\text{metal}) = \tfrac{1}{2} H_2(g)$$

According to Nozik (16) the energy available for electrolysis
when an n-type anode and metal cathode are used is:

$$E_{gap} - V_B - (E_c - E_F) = \frac{1}{nN_o} (\Delta G/F + \eta_a + \eta_c + iR + V_H) \qquad (8)$$

where the terms on the left-hand side refer to the semiconductor and its electronic equilibrium, whereas those on the
right refer to the two electrochemical charge-transfer reactions.
The minimum energy of light expected to be effective is
$h\nu = E_{gap} = E_c - E_v$. V_B is the amount of band bending, equal
to $E_v(\text{surface}) - E_v(\text{bulk})$ so that the left-hand side may be
simplified to $E_F - E_v(\text{surface})$. On the right-hand side, the
symbols n and N_o are the number of electrons transferred in a
single step of the reaction and Avogadro's number, respectively.
The first term in parentheses is the Gibbs free energy change in
the overall reaction, the second and third are the anodic
(equation (3)) and cathodic (equation (2)) overvoltages,
respectively, for oxygen and hydrogen evolution, the fourth is the
iR drop (resistive losses) term, which in the case of semiconductor electrodes includes losses from charge passage through
the solid phase as well as in solution, and the last term embraces
the potential drops across the solution double layers at the
anode and cathode. Using a semiconductor anode is likely to
drive up the overvoltage of the reaction occurring at that
electrode (η_a) compared with the overvoltage to be expected if a
metal were used instead. The double layer potential drops in
solution are likely to be small if moderately concentrated
solutions are employed, but the iR drop in the semiconductor
electrode may be inconveniently large unless heavily doped
material (of higher conductivity) is used.

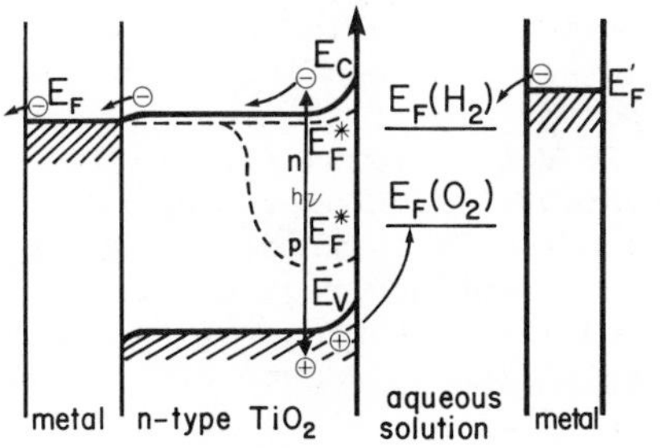

Academic Press

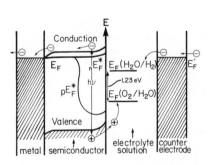

metal | n-type TiO₂ | aqueous solution | metal

Figure 6. Schematic showing energy correlations for photoassisted electrolysis of water using n-type TiO₂ as a photo-anode and a metal cathode. Symbols as in Figures 3, 4, and 5, except E_F is Fermi level for metal contact to TiO₂ and E_F' is higher Fermi level in metal cathode, polarized by an external source to a potential negative to the semiconductor anode. $E_{F(H_2)}$ and $E_{F(O_2)}$ are abbreviated forms for Fermi energies for redox systems of Figure 3 (13).

Academic Press

Figure 7. Schematic showing energy correlations for photoelectrolysis of water without external power source. An n-type semiconductor is depicted with a metal contact and connection through an external circuit to a catalytically active metal with $E_F' = E_F$. Other symbols as in Figures 5 and 6 (13).

Photo-electrolysis Without Applied Potential Using n-type Anode and Metal Cathode

Provided that light of sufficiently short wavelength is used, photo-assisted electrolysis should be relatively easy to achieve with a range of n-type semiconductors. TiO_2, SnO_2, WO_3, titanates and tantalates have been used, these materials having band gaps of 2.7 to 3.5 eV (20, 21, 22). The most interesting case, however, is that of achieving hydrogen and oxygen generation by action of light alone, i.e. without an applied potential. This has been achieved with the n-type anode / metal cathode configuration in only one case, that of $SrTiO_3$, which has a band gap of 3.2 eV but a greater amount of band bending at the interface than TiO_2, especially in strongly alkaline solution. The difference between TiO_2 and $SrTiO_3$ in band bending can be expressed by quoting the flat-band potentials of the two materials: 0 V for TiO_2 (23) and -0.25 V for $SrTiO_3$ (17, 20), both with reference to a standard hydrogen electrode. The consequence of this increased band bending is that the charge separation of electrons and holes in the surface region which is not possible for TiO_2 when illuminated in the absence of an applied voltage making E_F^* higher than E_F (as in Figure 6), becomes possible for $SrTiO_3$ (17, 18). This situation is illustrated in Figure 7. Strontium titanate is stable under conditions of use as an anode in aqueous acid or base. Photo-currents observed in one study at zero applied potential ranged from 0.5 mA to 1.5 mA cm^{-2} in 9.1 M NaOH, using a 200 W super-pressure mercury lamp focussed on the photo-anode (18).

Potassium tantalate, $KTaO_3$, has a similar degree of band bending (flat band potential of -0.2 V) but its band gap is even larger (3.5 eV) (20).

Photo-electrolysis With Semiconductor Cathodes

Logically, the use of a p-type cathode for evolving hydrogen and a metal anode is complementary to the systems just discussed. Difficulties arise in this case, however, because of the high overpotentials needed to drive hydrogen evolution on semi-conductors such as Si and GaAs. Gallium phosphide is a favourable instance, however, photo-evolution of hydrogen at illuminated p-type material occurring at a higher positive potential than expected (14), see e.g. Figure 10.

Nozik has proposed the combination of a p-type cathode with an n-type anode as a means of utilising light energy at both electrodes instead of at just one of them (16), i.e. a two-photon photo-electrolysis. In this proposal it should be possible to utilise semconductors with smaller band gaps than the 3.0 - 3.2 eV of TiO_2 and $SrTiO_3$, because it is only necessary that $E_F^*(n)$ be below the oxygen Fermi level and that $E_F^*(p)$ be above the hydrogen Fermi level. The energy conditions and the influence of light irradiation on them are illustrated in Figures 8 & 9.

The energy available for electrolysis (correcting an error of sign in (16)) when use is made of a p-type cathode and an n-type anode is:

$$E_{gap}(n) + E_{gap}(p) - V_B(n) + V_B(p) - E_c(n) + E_F(n) + E_v(p) - E_F(p)$$

which simplifies to $(E_F(n) - E_v(n, \text{surface})) + (E_c(p, \text{surface}) - E_F(p))$. This corresponds to the equilibrium situation depicted in Figure 8. This available energy is equated to the terms on the right-hand side of equation (8), so that

$$(E_F(n) - E_v(n, \text{surface})) + (E_c(p, \text{surface}) - E_F(p))$$

$$= \frac{1}{nN_o} (\Delta G/F + \eta_a + \eta_c + iR + V_H) \qquad (9)$$

Maximization of the available energy for electrolysis with a given semiconductor seems to require use of the most heavily doped material together with a minimal band bending consistent with efficient charge separation, so that in the n-semiconductor, $E_F(n)$ is as high as possible while $E_v(n, \text{surface})$ is as low as possible and in the p-semiconductor $E_F(p)$ is as low as possible and $E_c(p, \text{surface})$ as high as possible. Caution must be exercised, however, because a heavily doped semiconductor may become degenerate, i.e. metallic in behaviour and, as we have already noted, it is necessary to have suitable band bending to separate the charges.

Nozik's idea had already been tried by Yoneyama et al. (19). Figure 10 shows the photo-currents observed by these workers with a p-type GaP cathode and an n-type TiO_2 anode in H_2SO_4. Figures 11 and 12 show these workers' determinations of the flat band potentials of TiO_2 and GaP in the same electrolyte solution, both in the dark and with UV illumination. The positive potential shift upon illumination of TiO_2 and the negative potential shift at GaP are in the directions expected, but it was thought that the photo-currents of Figure 10 would have begun at the flat band potential for the illuminated case, whereas it actually commenced at both materials at potentials closer to the "dark" flat band value. This appears to be a favourable circumstance, perhaps, however, brought about by differing intensities of illumination in the two types of experiment. Suffice to say, the flat band potential is affected by the intensity of light, increased intensity producing greater potential shifts, as expected theoretically.

Figure 13 shows photo-currents observable at p-type GaP and n-type TiO_2 in NaOH, where again the flat band potentials did not quite coincide with the commencement of the photo-current. Currents in the dark for all of these cases were reported to be negligibly small.

232

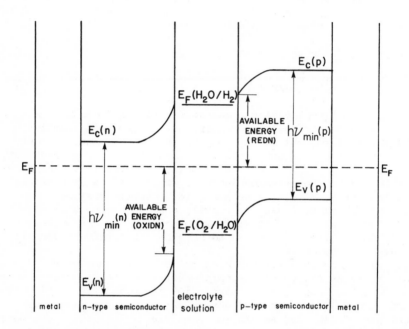

Figure 8. Schematic showing energy correlations at equilibrium for cell with two semiconductor electrodes in contact with aqueous solution and through an external circuit with each other. An n-type semiconductor anode and a p-type cathode are shown to left and right, respectively. In each case the minimum light energy to give rise to a photocurrent is indicated by $h\nu_{min}$ (n) and $h\nu_{min}$ (p), respectively. The energies available for oxidation and reduction are also indicated. $E_c(n)$ and E_v (n) are conduction and valence band edges for the n-type material and E_c (p) and E_v (p) are those for the p-type material. Other symbols as in Figure 7.

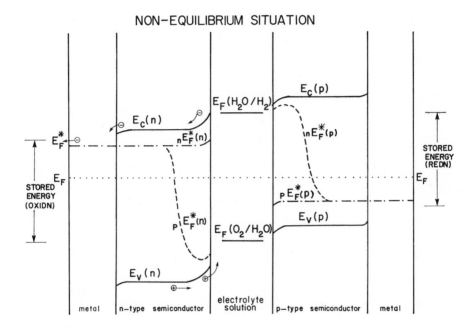

NON-EQUILIBRIUM SITUATION

Figure 9. Schematic showing energy correlations under conditions of illumination for the cell of Figure 8. Both electrodes are assumed to be illuminated and the general case of unequal band gaps is shown. The stored energy for oxidation is equal to $E_{F(O_2/H_2O)} - {}_nE_F^*$ (n, surface) while that for reduction is $E_{F(H_2O/H_2)} - {}_nE_F^*$ (p, surface). E_F denotes the Fermi level as it was at equilibrium in the dark in Figure 8, while E_F^* and E_F^{**} are the Fermi levels in the two metals when the semiconductors are both illuminated. Other symbols as in Figures 7 and 8 (13).

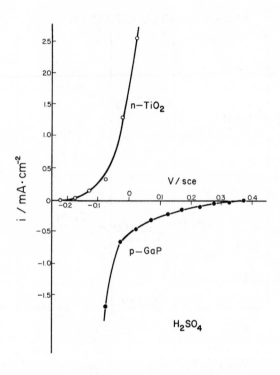

Figure 10. Cathodic polarization curve of p-type GaP and anodic polarization curve of n-type TiO₂ in 0.5M H₂SO₄ with illumination by an ultrahigh-pressure mercury arc lamp (19).

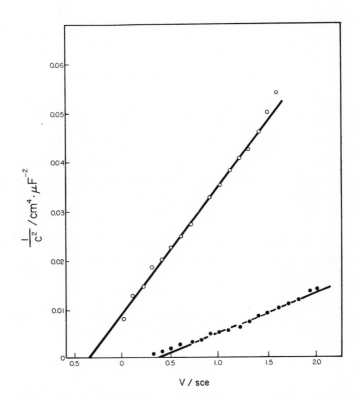

Figure 11.   Mott–Schottky plots of reciprocal square of differential capacitance of n-type TiO$_2$ electrode in 0.5M H$_2$SO$_4$ vs. electrode potential. (○) In the dark; (○) under illumination as in Figure 10. Intercept at C = ∞ gives the value of the flat-band potential (19).

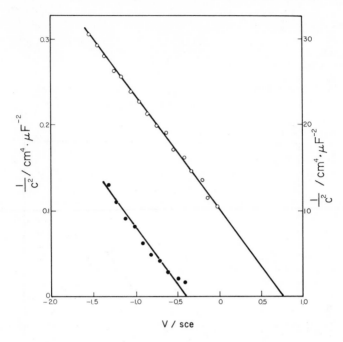

Figure 12. Mott–Schottky plots as in Figure 11 but for p-type GaP in 0.5M H₂SO₄. Symbols have the same meaning as in Figure 11 (19).

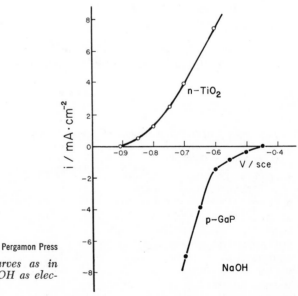

Figure 13. Polarization curves as in Figure 10, but with 1M NaOH as electrolyte (19)

Photo-assisted Electrolysis in Simulated Sunlight

Figure 14 shows the spectral responses for TiO_2 and for GaP with, for comparison, the energy distribution in sunlight. The band gaps of the two materials are also indicated. As is evident from this diagram the light energy range in which TiO_2 functions is well away from the broad maximum in the solar energy incident upon the Earth (at 500 to 900 nm for air mass 2 (24)). Even for GaP the band gap indicated at 2.38 eV is an indirect one, the maximum for this material occurring closer to the direct gap at 2.77 eV. Thus both materials may be expected to function less effectively than desirable in utilization of solar energy. Nevertheless, they have been combined by Nozik (16) into a cell with photo-effects at both electrodes as shown in Figure 15. With this cell, hydrogen and oxygen were produced for the first time from simulated sunlight with zero applied potential.

In Figure 16 is shown the photo-current / potential relation for an n-type GaP anode and a p-type GaP cathode. The current at zero applied potential is larger in this case, but this is because the anodic reaction is dissolution of the GaP and not oxygen evolution.

Future Prospects

The prospects for an economic process for photo-electrolytic H_2 and O_2 evolution are uncertain. It is necessary to find materials with smaller band gaps than those used successfully in ultra violet light as anodes and to find some alternative materials, also with smaller band gaps, for use as cathodes. The double photo-effect approach of Nozik appears to be the most promising and GaP might prove adequate as a cathode if no better material becomes available. The recently reported (25) extension of spectral response of p-GaP to beyond 1000 nm by electro-deposited monolayers of Ag, Pd or Au is an encouraging development. Apparently, hydrogen uptake is implicated in the phenomenon. For anodic O_2 evolution, oxides appear the most promising but some with suitable band gaps, e.g. $\alpha-Fe_2O_3$ (2.2 eV (27)) have unsuitable band bending (+ 0.7 eV) (20, 28. 29), necessitating a large applied potential and thus defeating the original purpose. The increase in photo-oxidation efficiency at polycrystalline $\alpha-Fe_2O_3$ in the presence of chelating agents such as citrate or EDTA (30) may be of advantage.

Other problems are those of materials costs and material processing costs, e.g. whether flat single crystals have to be used. Finally there is the question of spatial requirements, both for interception of solar radiation and for provision of a suitably large electrode area to give useful rates of H_2 generation. One may wonder, with Gerischer (13), whether it might not be better to generate electricity by photo-voltaic means, using this to electrolyse water in a conventional fashion.

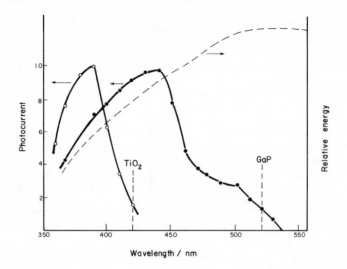

Pergamon Press

Figure 14. Spectral responses of n-*type* TiO_2 *(○) and of* p-*type* GaP *(○) in 0.5M* H_2SO_4 *while polarized at* $-1.0\,V$ *vs. SCE (GaP) or at* $+1.0\,V$ *vs. SCE (TiO₂). The band edges (gaps) for each material are indicated by vertical dashed lines. The solar energy distribution (———) is also shown. Photoresponses and solar energy distribution are in arbitrary units (19).*

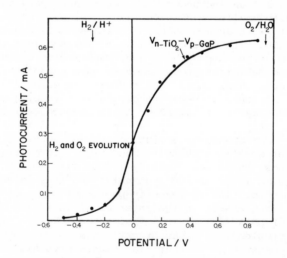

American Institute of Physics

Figure 15. Photocurrent/cell potential difference for n-*type* TiO_2 *anode and* p-*type* GaP *cathode in 0.1M* H_2SO_4, *illuminated by ca. 100 mW cm⁻² of simulated sunlight from a xenon lamp. The* H_2/H^+ *and* O_2/H_2O *reduction potentials are indicated. Hydrogen and oxygen evolution were obtainable at zero potential difference as indicated (16).*

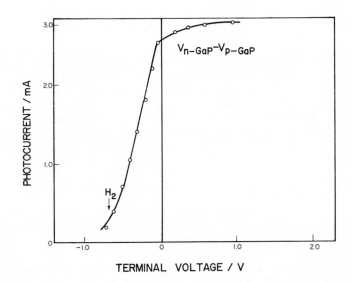

Figure 16. Photocurrent/cell potential difference for n-type GaP anode and p-type GaP cathode in 0.1M H_2SO_4, illuminated as in Figure 15. Hydrogen evolution occurred at the GaP cathode without visible degradation, but in this cell the anodic reaction is oxidation of P^{3-} to H_3PO_3 and dissolution of the anode material (16).

However, the approaches discussed here may well have special
advantages for fuel generation in remote unattended locations,
e.g. in Canada's North.

Abstract

The advantages of hydrogen storage combined with the
generation of electricity by solar radiation make the concept of
photo-electrolysis of water an attractive one. Photo-effects
have been observed both at oxygen evolving anodes and at hydrogen
evolving cathodes, both being of use in practical photoelectrolysis when the incident radiation is ultraviolet. The
requirements are of semiconductor anodes or cathodes which do not
decompose in the electrolyte and which are electrocatalysts for
oxygen or hydrogen evolution from water. The band gaps of the
semiconductors should be reasonably matched to the solar spectrum
for high efficiency and a suitable degree of bending of the bands
at the surface is desirable. Materials that have been used as
anodes include TiO_2, various titanates and tantalates, WO_3 and
Fe_2O_3. GaP stands essentially alone as a cathode material.
Prospects for the future are surveyed.

Literature Cited

1.   Kuhn, A.T. (editor), "Industrial Electrochemical Processes",
     Elsevier, (1971).
2.   Smith, D.H., CJB Developments, Portsmouth, Chemical Society
     Conference on Electrolytic Production of Hydrogen, City
     University, London, 25th February 1975.
3.   Nuttall, L.J. and Titterington, W.A., General Electric Co.,
     Wilmington, Mass., also presented at meeting of Ref. 2.
4.   Brodsky, A.M. and Pleskov, Yu.V., in Davison, S.G. (editor),
     "Progress in Surface Science, (1972), 2, 1.
5.   Brattain, W.H. and Garrett, C.G.B., Bell. Syst. Tech. J.,
     (1955), 34, 129.
6.   Gerischer, H., Anales de Fisica y Quimica, (1960), 56, 535.
7.   Gobrecht, H., Kuhnkies, R. and Tausend, A., Z. Elektrochem.,
     (1959), 63, 541.
8.   Smith, R.A., "Semiconductors", p. 383, Cambridge Univ. Press,
     1964.
9.   Gerischer, H., in Delahay, P., (editor), "Advances in
     Electrochemistry and Electrochemical Engineering", Vol. 1,
     p. 139, Wiley-Interscience, New York, (1961).
10. Fujishima, A., and Honda, K., Nature (London), (1972), 238,
     37.
11. Ohnishi, T., Nakato, Y. and Tsubomura, H., Ber. Bunsenges.
     Phys. Chem., (1975), 79, 523.
12. Lohmann, F., Z. für Naturforschung, (1967), 22a, 843.
13. Gerischer, H., in Bolton, J.R., (editor), "Solar Energy and
     Fuels", p. 77, Academic Press, New York, (1977).

14. Memming, R., and Schwandt, G., Electrochim. Acta, (1968), 13, 1299.
15. Meek, R.L. and Schumaker, N.E., J. Electrochem. Soc., (1972), 119, 1148.
16. Nozik, A.J., Appl. Phys. Lett., (1976), 29, 150.
17. Mavroides, G., Kafalas, J.A. and Kolesar, D.F., Appl. Phys. Lett., (1976), 28, 241.
18. Wrighton, M.S., Ellis, A.B., Wolczanski, P.T., Morse, D.L., Abrahamson, H.B. and Ginley, D.S., J. Amer. Chem. Soc., (1976), 98, 2774.
19. Yoneyama, H., Sakamoto, H. and Tamura, H., Electrochim. Acta, (1975), 20, 341.
20. Mavroides, G., Canadian Association of Physicists Informal Meeting on Solar Energy, Montreal, October 1977.
21. Bolts, J.M. and Wrighton, M.S., J. Phys. Chem., (1976), 80, 2641.
22. Hodes, G., Cahen, D. and Manassen, J., Nature (London), (1976), 260, 312.
23. Möllers, F., Tolle, H.J. and Memming, R., J. Electrochem. Soc., (1974), 121, 1160.
24. Archer, M.D., J. Appl. Electrochem., (1975), 5, 17.
25. Yoneyama, H., Mayumi, S. and Tamura, H., J. Electrochem. Soc., (1978), 125, 68.
26. Hardee, K.L. and Bard, A.J., J. Electrochem. Soc., (1977), 124, 215.
27. Strehlow, W.H. and Cook, E.L., J. Phys. Chem. Ref. Data, 1973, 2, 163.
28. Quinn, R.K., Naseby, R.D. and Baughman, R.J., Mater. Res. Bull., (1976), 11, 1011.
29. Hardee, K.L. and Bard, A.J., J. Electrochem. Soc., (1976), 123, 1024.
30. Kennedy, J.H. and Frese, K.W., Jr., J. Electrochem. Soc., (1978), 125, 709.

RECEIVED September 25, 1978.

16

Electrochemically Codeposited Large-Area Photoelectrodes for Converting Sunlight to Electrical Energy

B. LIONEL FUNT, MARZIO LEBAN, and ALDEN SHERWOOD

Department of Chemistry, Simon Fraser University, Burnaby,
British Columbia, Canada V5A 1S6

Photoelectrochemical cells based on semiconductor photoelectrodes are potential candidates for low cost, large area, conversion devices. The basic starting materials are inexpensive and photoelectrodes of large area can be produced by electrodeposition.

In this paper we describe the construction and performance of a photoelectrochemical cell of 100 cm^2 photoelectrode area and indicate some of the parameters which are affected by the scale-up from typical electrodes of under 1 cm^2.

The photoelectrochemical converter is composed of a semiconductor in contact with a solution containing a suitable redox couple. CdSe with a band gap (E_g = 1.7 eV) utilizes a large part of the solar spectrum. These electrodes can be prepared by cathodic codeposition [1] on Ni substrates. Until recently such photoelectrochemical converters were impractical because the action of light on the surface of the photoelectrode produced dissolution.

$$CdSe \xrightarrow{h\nu} Cd^{+2}_{(aq)} + Se_{(s)} + 2e^-$$

However, Wrighton et al [2-6] showed that the photoelectrodes could be stabilized in a polysulfide medium. The recognition that these devices could show high stability, large area, and low cost, spurred on further investigation of the factors involved in producing working devices.

Experimental

a) <u>Production of Photoelectrodes</u>. CdSe photoelectrodes were prepared by codeposition of Cd and Se on nickel foil substrates. This electrochemical method is a slight variation of a technique devised by G. Hodes et al [1], to whom we are indebted for supplying additional details of their processes.

The nickel foil and a cadmium electrode were immersed in a stirred solution which was 0.5 M in CdSO$_4$ and 1 M in NH$_4$Cl. Cadmium and nickel electrodes were shorted through an ammeter. This resulted in a rapidly decaying current which stabilized at about

0.1 mA after a few minutes. SeO_2 was added, in solid form, (~20 mg at a time) with a spatula and produced currents of up to 40 mA during codeposition of CdSe. Deposition times of up to 45 minutes were employed. Repeated additions of SeO_2 prevented the current from falling below 20 mA.

Finally, the deposits were heat treated at 400°C for 5 minutes under a N_2 atmosphere. X-ray fluorescence data confirmed the presence of Cd and Se in the deposits with no other detectable impurities. CdSe photoelectrodes, 10 cm^2 in area (2 x 5 cm) were prepared in this manner with a success rate of about 50%.

b) Counterelectrodes. Union Carbide "grafoil" was used as the counterelectrode material. The .010" thick graphite foil was found to be convenient to use and to perform as well as graphite rods.

c) Cell Design. The prototype photoelectrochemical cell is shown in Figure 1. Ten 10 cm^2 CdSe photoelectrodes make up the 100 cm^2 cell. Photoelectrodes were mounted in tandem on plastic strips. Six counterelectrodes, each on a plastic support, were mounted perpendicular to and in between the photoelectrodes. See Figure 2 for the unmounted cell. Each electrode has its own external connection, allowing individual testing. The entire cell, including the cylindrical lens panel shown in Figure 1, was constructed of plexiglass. This panel, intended to focus light on the photoelectrodes, was constructed by gluing together five cylindrical lenses each with a 4 cm focal length.

d) Light Source. The light source was a 200 w, 24 v, EJL Quartz Halogen lamp with a maximum intensity at about 6500 Å. Wavelength vs. Irradiance was obtained with a Jarrell Ash Model 82-410 Spectrophotometer and a Model 65 A YSI Kettering Radiometer.

Theory

The photoelectrochemical cell consists of a CdSe photoelectrode immersed in an aqueous solution which is 1 M in NaOH, Na_2S, and S. When the CdSe photoelectrode is illuminated, electrons are generated which may flow through the external circuit to the counterelectrode. The CdSe photoelectrode acquires a negative charge and the counterelectrode a positive charge.

Upon immersion of the CdSe semiconductor into the electrolyte, electron exchange at the interface occurs until equilibrium is attained. At equilibrium, the Fermi level of the semiconductor is adjusted by the presence of a space charge layer at the semiconductor surface. This layer is due to the difference between the Fermi level of the semiconductor and the Fermi level of the electrolyte which is measured at the redox couple (7). The potential drop at the space charge layer and the amount of band bending also depend on the degree of Fermi level mismatch at the semiconductor-

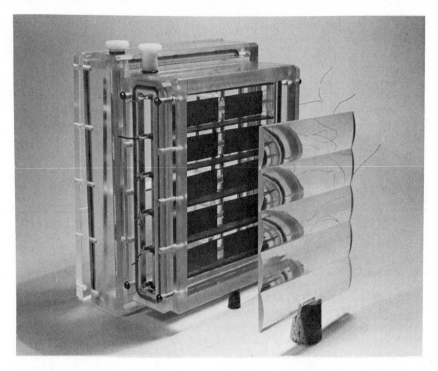

Figure 1.   The 100-cm² CdSe photoelectrochemical cell

Figure 2.   Unmounted cell; photoelectrodes on the right, counterelectrodes on the left

electrolyte interface. Figure 3 illustrates band bending for the top of the valence band (VB) and the bottom of the conduction band (CB). The horizontal coordinate represents distance from the interface, while the vertical represents energy. The direction of band bending shown is typical of an n-type semiconductor such as CdSe. Opposite band bending would occur for a p-type semiconductor (3). The Fermi level of the redox couple, E_F, referred to the vacuum level, is given by (9):

$$E_F = Constant-E^o$$

where E^o is the equilibrium redox potential measured against the standard hydrogen electrode, and the value of the constant is 4.5 eV (15,16).

Band bending at the semiconductor surface causes a depletion of the majority carriers (electrons for n-type CdSe) underneath the surface (depletion layer). Formation of a depletion layer gives rise to a system equivalent to a Schottky barrier between a metal and a semiconductor.

In the dark, the system is at equilibrium and no current or potential is observed. When the semiconductor absorbs a photon with $h\nu \geq$ band gap energy, an electron from the valence band may be promoted to the conduction band. A hole (positive charge) is created in the valence band (8). When electron hole pair formation occurs within the depletion region, the minority charge carriers (holes) are swept to the semiconductor electrolyte interface. At the same time, electrons move into the semiconductor bulk (9). Electron hole recombination may occur before holes reach the surface or electrons reach the semiconductor bulk resulting in the formation of heat, (Electron + hole = heat). Motion of electrons and holes within the depletion region is due to the electric field in the space charge layer. Holes (positive charges in the valence band of Figure 3) that reach the semiconductor electrolyte interface may undergo electron exchange with the redox couple in the electrolyte (10). Here, the redox couple is S/S^{-2}. Holes behave as oxidizing agents and may be neutralized by electrons of the sulfide or polysulfide species in solution (10). At the counterelectrode, the opposite reaction occurs (formation of sulfide or polysulfide) resulting in no net chemical change of the electrolyte. The S/S^{-2} redox couple not only quenches dissolution of the photoelectrode and completes the internal circuit of the photoelectrochemical cell, but also provides a mechanism for extracting the energy momentarily stored at the photoelectrode. In directly converting sunlight to electrical energy, the energy (obtained from light) absorbed by the semiconductor photoelectrode is stored as photogenerated electron hole pairs. The lifetime of these pairs is long enough to allow the redox reactions to compete with electron hole recombination. These properties of the semiconductor photoelectrode render photoelectrochemical cells useful as solar energy converters.

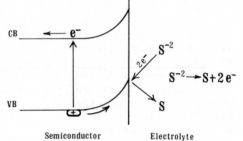

Figure 3. The semiconductor electrolyte interface; CB, conduction band; VB, valence band; positive charges represent holes

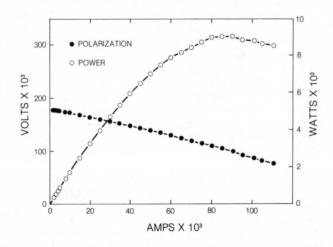

Figure 4. Voltage–current–power characteristics; light intensity 92.5 mW/cm²

Performance

Each photoelectrode was tested prior to mounting. The maximum and minimum photocurrents and potentials observed for the 10 electrodes were 80 and 30 mA and 450 and 300 mV, respectively. The effect of constant illumination for 20 hours on a photoelectrode was to reduce the short current photocurrent by 15%. However the original output was almost restored when the electrode was placed in a fresh solution. The long term stability reported by Hodes is encouraging (11).

Semiconductor deposits thicker than the depletion layer ($\sim 10^{-6}$ cm) are required for optimum performance. The CdSe deposits were estimated to range up to 10^{-4} cm. The deposits were not doped chemically, but the heat treatment vaporizes Se leaving excess Cd which may act as a dopant. A marked increase in short circuit photocurrents was observed after heat treatments.

The resistance between the semiconductor deposit and the nickel substrate was less than 2 ohm, indicating a good contact between deposit and substrate.

Each photoelectrode was retested after mounting and current voltage plots obtained. All photoelectrodes decreased in performance and one became essentially inoperative. This is attributed to excessive handling during mounting which resulted in abrasion and deterioration of the deposit. The maximum power output of the individual mounted electrodes in the assembly varied from 5.70 mW to 0.13 mW. The maximum power obtained from the assembled converter was 27 mW. Under short circuit conditions the maximum power at 90.2 mA and 100 mV was 9.0 mW. The light intensity incident at the photocell was 92.5 mW/cm^2.

Some additional factors should be considered in placing these results in perspective. Although the counterelectrode surface area was almost twice that of the photoelectrode, this may not have been adequate to minimize the overpotential. Any overpotential will decrease output voltages and efficiencies under a load (13,14). Low overpotential catalytic counterelectrodes may offer a solution to this problem (13). It is also possible that inhomogeneous illumination of the entire 100 cm^2 may have contributed to the reduced power output.

Scale-Up Considerations

Intermediate sized photoelectrodes of 25 cm^2 (5 x 5 cm) were prepared with partial success. Uneven deposits were formed as indicated by patches of discoloration. Initial photocurrents of 75 mA were observed which quickly deteriorated to 1.5 mA. The unsuitability of "beaker type" preparation conditions is evident for large area deposition. Large area photoelectrode production will require a deposition technique with controlled potential or current. Experiments have shown that with controlled potential CdSe photoelectrodes may be prepared in a more reproducible manner.

Photoelectrodes formed in this way will be incorporated in our next 100 cm^2 photoelectrochemical cell.

Certain conditions of cell design must be met. In order to decrease internal cell resistance the counterelectrode should be as close as possible to the photoelectrode, but must not obstruct the light. The modular counterelectrode design, Figure 1, meets this requirement but limits the actual size of the photoelectrode. The ideal counterelectrode would be transparent to light and would be deposited on the inside of the cell window. This geometry allows cell designs of single large area photoelectrodes with counterelectrode proximity. SnO_2 is a candidate for such a counterelectrode. The path length of light in the electrolyte should be limited since absorption occurs in the near UV with present electrolytes.

An objective of this work was the assessment of the factors relevant to scale-up to useful power outputs. The preliminary results thus far obtained have been encouraging. The preparation of larger electrodes has proved to be simple and there are no apparent limitations to the preparation of much large electrodes by these techniques. The costs of the starting materials, Cd and Se are quite low and relatively small quantities are required in the preparation of a photoactive surface. Overall material costs are therefore quite modest. The stability of the photoelectrode is still under investigation. Hodes (11) reported good stability for a number of light-dark cycles over several weeks. Our measurements indicate relative stability of individual electrodes. However there has been a significant deterioration in the performance of the assembled 100 cm^2 converter; and this is attributed in part to the effects of handling and mounting.

The reduction in overall efficiency on scale-up deserves further design and performance consideration. The net power output was lower than expected from the summed performance of the individual units and the optical, solid state or electrochemical factors which contribute to this reduction in performance need to be classified.

The reported performances (4,5) and our own observation indicate that individual photoelectrodes with a conversion efficiency of sunlight to electrical energy of 2% may be prepared. A realistic objective is to retain an efficiency of ≈1% in a large working converter with an output of ≈10 w/m^2.

Some of the considerations for design of larger converters are:

(1) Production of reproducible photoelectrodes, doped electrochemically, with reasonable outputs (i.e., 3 mA/cm^2 at 350 mV).

(2) Use of counterelectrodes with low polarization and low overpotentials.

(3) Optimization of light transmission through the electrolyte by reduction of sulfur concentration and path length.

(4) Minimization of cell resistance.

(5) Investigation of alternative redox couples for

stabilization of photoelectrode, but with better optical transmission in solution.

In our opinion preliminary results thus far obtained warrant further intensive investigation of photoelectrochemical converters as potentially important and useful devices.

Acknowledgement

A major part of this work was performed under DSS Contract #8AU77-00433.

We are indebted to Dr. D. Snelling for the performance measurements of current voltage and power shown in Figure 4, and for other relevant data.

Literature Cited

1. Hodes, G., private communication.
2. Ellis, A.B., Kaiser, S.W., and Wrighton, M.S., J. Am. Chem. Soc. (1976) 98, 1635.
3. Williams, R., J. Chem. Phys. (1960) 32, 1505.
4. Ellis, A.B., Kaiser, S.W., and Wrighton, M.S., J. Am. Chem. Soc. (1976) 98, 6855.
5. Ellis, A.B., Kaiser, S.W., Bolts, J.M., and Wrighton, M.S., J. Am. Chem. Soc. (1977) 99, 2839.
6. Ellis, A.B., Kaiser, S.W., and Wrighton, M.S., J. Am. Chem. Soc. (1976) 98, 6418.
7. Gerischer, H., "Physical Chemistry: An Advanced Treatise", Vol. 9A, Chapter 5 (1970), Eyring, H., Henderson, D., and Jost, W., Ed., Academic Press, New York, N.Y.
8. Gerischer, H., J. Electrochem. Soc. (1966) 113, 1174.
9. Gerischer, H., J. Electrochem. Soc. (1975) 58, 263.
10. Gerischer, H., Surface Sci. (1969) 13, 265.
11. Hodes, G., Manassen, J., and Cahen, D., Nature (1976) 261, 403.
12. Snelling, D., Defence Research Establishment, Ottawa. Private communication.
13. Hodes, G., Manassen, J., and Cahen, D., J. Appl. Electrochem. (1977) 7, 181.
14. Manassen, J., Hodes, G., and Cahen, D., J. Electrochem. Soc. (1977) 124, 532.
15. Reiss, H., J. Electrochem. Soc. (1978) 125, 937.
16. Lohmann, F., Z. Naturforsch. (1967) 22a, 843.

RECEIVED August 22, 1978.

Electricity Production and Storage

Electrochemical Energy Storage Systems

On the Selection of Electrolytes for High-Energy Density Storage Batteries

E. J. CASEY and M. A. KLOCHKO

Defense Research Establishment, Ottawa, Ontario, Canada K1A 0Z4

Relevance of Electrical Energy Storage. The importance of energy storage seems to derive from the versatility it offers to an energy system. The controlled storage, transformation and use of stored energy permits us to conserve any excess energy from the source, and to disburse it later, presumably wisely and according to need. Possible transformations which storage permits are: (a) controlled time-phasing, (b) energy-form changes, and (c) controlled disbursement into loads (Figure 1).

In the context of this Symposium entitled Chemistry for Energy, the task set for the authors by the organizers was to draw a perspective on electrical energy storage -- not only to summarize the state-of-the-art and indicate why performance is limited, but also to draw attention to chemical problems, by way of example, which might capture the participants' attention.

The renewed activity and support of research and development of this field seems to have arisen as a result of the assumptions which follow:
(a) Research, if to be supportable politically, must be demonstrably relevant to the solution of some national problem.
(b) Energy-storage, -transport and -distribution are such problems.
(c) Renewable resources will be much more easily harnessed and be more widely used once better methods of conversion and storage become available.
Storage of electrical energy holds a special place. The further assumptions here seem to be:
(a) Improved electrical energy storage will relieve the technical and economic pressures on the total energy system, by simplifying the (centralized and localized) storage and redistribution of electricity, as well as by enabling the introduction of electric vehicles on a massive scale.

----

*Issued as DREO Report No. 785.

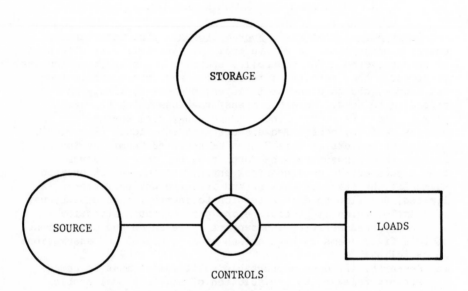

Figure 1. Energy system with storage. Possible transformations include: controlled time phasing, energy-form changes, and controlled disbursement to loads.

(b) Increased research activity in the electrochemical field will result in better batteries and in alternate electrical-energy-storage systems not even yet imagined.

Each of these assumptions can be challenged, but they are difficult to prove false, or even unjustified. Analysts differ in their judgements about what level of effort should be funded by governments. The U.S. made a conscious decision in 1974 to invest heavily in energy R & D; Canada was wrestling with other priorities. Electrical-energy storage R & D has received substantial fall-out in the U.S.A., and the multi-million dollar hydrogen, battery, fuel cell, and electrolyzer programs have flowered, along with the more spectacular nuclear, shale-oil, coal gasification and other major efforts. In Canada, tar sands and oil exploration have received added stimulus. The nuclear R & D program assumes 2/3 of the federal energy R & D budget. The electrical-energy-storage and hydrogen programs have received some new stimulus.

The interests of our Colleagues within the Energy Conversion Division of Defence Research Establishment Ottawa include energy related problems resolvable by batteries, fuel cells, engines, windmills, thermoelectric generators, etc.; and include modelling of energy conservation methods as well. At the end of this paper we give an example of a modelling study prepared in answer to the question: "Energy storage? First of all, who wants it within our own organization? And why?

Reference Points for this Review. Rather than prepare a tutorial on the subject, we have taken the invitation as an opportunity to try to review the storage-battery field from a new point of view, one not treated in the literature. We focus on the electrolyte and try to uncover untried but potentially useful new storage battery systems based on light metals.

In considering the selection of anodes for high energy density (HED) storage (or secondary) batteries (SB), we note that there are some 19 metals whose free-energy density (TED) of reaction with oxidants such as O_2, Cl_2, and F_2 are higher than those of Zn with the same oxidants. Most of these metals react violently with water. The remainder are passivated by water. Therefore other electrolytes must be considered for these metals, based on non-aqueous, molten salt, or solid-state ionic conductors. Much experimental work has been carried out during the last two decades on primary and secondary batteries based on anhydrous electrolytes, aimed at utilization of the active metals.

TABLE I.   Relevant Properties of Some Active Anodic Metals

Metal	Atomic No.	g.-Equiv.	Abundance ppm	$E°$ 25°	TED, Wh/kg M+O$_2$	M+Cl$_2$	M+S
Li	3	6.94	65	-3.04	5536	2680	3020
Na	11	22.99	28300	-2.71	1873	1954	1329
Ca	20	20.05	36300	-2.87	3148	1991	1850
Mg	12	12.15	29900	-2.37	4158	1874	1712
Al	13	9.00	81300	-1.66	4609	1450	1320
Ti	22	23.95	4400	-1.61	2305	1116	1070
Zn	30	32.68	132	-0.76	1211	848	579

Six of the 19 active metals can be singled out as the most promising ones for HED SBs: Li, Na, Ca, Mg, Al and Ti. Being located in the upper left-hand corner of the periodic table (even Ti is close) these metals possess HED's in their reactions with many other substances and have high potentials and low equivalent weights. Four of them, viz. Li, Na, Ca and Ti, have begun to attract attention of the researchers during the last decade. The main obstacle for the introduction of Mg, Al and Ti into SBs is their passivation. By contrast, the use of the other 3 metals, Li, Na and Ca, is hindered by the rate of their reaction with water. However, the high energy of the reactions of all these six metals with the common oxidants, and their abundance in the earth's crust are facts which quicken one's interest in their possible use. Some of the properties of these metals are given in Table I. They can be compared with zinc, which is currently the most widely used metal in primary batteries and is also the working anode in the Ni-Zn and Ag-Zn SBs.

Little practical success has been achieved up to now in the design and construction of SBs based on light metals, despite extensive work done in that direction. The only new SBs which are today under test for practical large-scale use are the ones based on the reversible reaction Na + S, with a solid electrolyte of the type Na$_2$O(Al$_2$O$_3$)$_{11}$, and the SB based on the reversible reaction of Li + FeS$_x$ in a molten chloride electrolyte.

To prepare this report we first composed a referenced account of research work on electrolytes which seemed to be pertinent to the development of truly advanced rechargeable battery systems. We note some opportunities which seem to exist with (a) dry organic and inorganic systems, with (b) solid- electrolytes and with (c) molten-salt electrolyte systems. Then we suggest

(d) that the slightly solvated low-melting salt systems may become attractive in the future. The survey part therefore naturally falls into four parts. In the final sections we summarize a recent analysis of what to expect from an electrical-energy storage system if introduced into small isolated communities in Canada's North, and then we give our current perception of the present points of view of Canadian manufacturers and industrialists to whom electrical-energy storage is vital.

First, as a reference point some basic battery lore may not be out of place. Good battery electrolytes are good ionic conductors. These could be aqueous solutions, molten salts, solid electrolytes (inorganic, organic), or solutions of salts in dry organic or inorganic solvents. Not only must the electrolyte be a good conductor, it must be resistant to anodic oxidation and cathodic reduction. Nor can it be unmanageably corrosive to electrode materials, case, or seals. Ideally it should be innocuous or non-poisonous as well. It should be stable thermally over a useable range of temperature. And it should have a low freezing point, permitting operation at low tempera-tures. One values safety and longevity, and low cost as well. Many intriguing redox systems have failed to be accepted because of poor prospects on safety or cost.

Sulfuric acid is an acceptable electrolyte -- a good ionic conductor, affordable, readily available, low freezing point if concentrated, etc. -- the basis of the Pb/PbO_2 storage battery system. The reaction product, $PbSO_4$, is highly insoluble and therefore tends to remain in place, on the plates, during cycling. Table II gives some background information on the lead acid system, with which everyone is more or less familiar. Although heavy, it is reliable, and forms the backbone of the storage battery industry world-wide. Its energy-density of only 26 Wh/kg, however, stimulates one to look elsewhere for systems potentially capable of much higher energy density. We look for chemical reactions with high theoretical energy density, strong electrolytes, as circumscribed above, electrochemical revers-ibility of insoluble reactants and products, low polarizations at both electrodes, and high turnaround efficiency: energy out divided by energy in.

The theoretical understanding of strong electrolytes based on molecular and cluster theories is not well enough advanced to permit a priori selection of electrolytes for high-performance battery systems which could incorporate the light, active anodes. The concentration of void spaces in pure and in two-component solvents do seem to correlate with conductivity and freezing point, but the curious complexing which occurs in molten salts, and which seems inextricably related to the morphology of the redeposited metal at the anode, is simply not well enough understood that one can predict with any confidence. We are reduced ultimately to making judicious choices and trying them out.

TABLE II.   Lead Acid Storage Battery (SB) : ED vs TED

$$(-) \quad Pb \; // \; H_2SO_{4_{aq.}} \; // \; PbO_2 \quad (+)$$

(a) $Pb + PbO_2 + 2 \; H_2SO_4 \quad \xrightarrow[\text{chge}]{\text{disch.}} \quad 2 \; PbSO_4 + 2 \; H_2O$

$$E_{rev.} \sim 2.0V$$

(b) $2 \; \dfrac{equiv.}{mole} \; \times \; 96,500 \; \dfrac{cou.}{equiv.} \; \times \; 2.0 \; \text{volts} = 4 \times 10^7 \; \text{jou/mole}$

* theor. energy density (TED)
      200 Wh/kg ($\sim$ 100 Wh/lb) of reactants only
* practical energy density (ED)
      26 Wh/kg ($\sim$ 12 Wh/lb) of battery (all components)

(c)   Practical energy density decreases with power withdrawn, and decreases with decreasing temperature -- and age.

(d)   safe .... reliable .... affordable

      heavy .... slow charge acceptance at low temperatures

"Nonaqueous" Electrolytes

Many organic and inorganic compounds which are liquid at ambient temperatures have been considered as solvents for light metal salts, to determine whether the solutions so formed could be used as electrolytes in electrochemical cells (1,2). Reversibility at both electrodes is required of electrochemical processes in cells intended for secondary batteries. This condition is a difficult one. Only solvents can be used from which the anode metal can be electrochemically deposited. On account of the rapid or even violent reaction between the alkali metals and water, nonaqueous aprotic solvents have been sought as the basis for electrolytes for power cells using these metals as anodes. However, the salts of these metals, especially of the MX type, where M is the metal and X the halogen, are generally not soluble in aprotic liquids. "Complex solubilization" can be employed: in the presence of Al halides the MX salts are fairly soluble in many aprotic solvents because of complex formation. Applying this method for the preparation of solutions of alkali metal salts in nitrobenzene, for example, one of the authors some years ago succeeded in depositing Li, Na, K, Rb and Cs on the

cathode at ambient temperature (3). Very recently Tobias used
this method again and described new results (4).

Primary Battery Development. Many publications (5-15) and
presentations have occurred in the '70's on PRIMARY cells based
on dry organic or inorganic electrolytes. HED's have been
achieved. Low-temperature output has been achieved with some.
Table III lists some of the more popular solvents. Preferred
electrolytes include $POCl_3$ + $LiPF_6$ or $LiBCl_3$, and $SOCl_2$ + $LiAlCl_4$
or $LiAsF_6$.

Rechargeability has been claimed in press releases, and in
one paper on Li/TiS_2 (small (4%) depth of discharge (dod)). But
high cycle life at reasonable dod's has NOT been achieved and
confirmed yet. The problem is that the lithium, calcium or
sodium will not redeposit (recharge) at operationally useful
currents. Many labs are working on this problem, including our
own. Perhaps our contemporaries may soon be selling a cell, or
disclosing how to make a HED cell, capable of 1000, say, 80% dod
cycles, the package (hopefully) containing at least 150 Wh/kg.

Details on Li primary cells based on three inorganic solvents
-- $POCl_3$, $SOCl_2$ (thionylchloride) and SO_2Cl_2 (sulfurylchloride)
-- are given in (6). In a later paper (8) cells based on the
$Li/SOCl_2$ + $LiAlCl_4$/C system, with E = 3.90V and ED theor. = 1875
Wh/kg, which had been selected for development as primary
batteries, were described. The work of Marincic (11) and Behl
(13) has contributed substantially to the development of such
primary cell systems.

Some other inorganic solvents were tried with Li anodes as
well (16-18).

A great many publications, articles, reports and patents on
the use of organic solvents for HED cells have appeared during
the last 15 years. Most of these publications (perhaps over 90%
of them) are concerned with Li cells and were recently reviewed
in a paper which includes in its 169 references the literature up
to 1974 (19). Some data used in this preliminary work can be
found in (1-4) and in (19-21), and in the publications cited in
those references. As a result of the efforts spent during the
last decade on primary nonaqueous batteries, at least some of
these systems are now achieving the commercial stage. The
attractive feature of such batteries is the high theoretical TED
of their redox couples, accompanied by high OCV. In a table
presented in Gilman's overview of the primary Li battery program,
which is based mostly on propylene carbonate (PC) solutions with
a Li anode and various cathodes, the OCV for a $(C_4F)_n$ cathode was
given as 5.23V, the theoretical ED being 2020 Wh/kg and the
experimental ED = 1228 Wh/kg. For the $(CF)_n$ cathode the figures
are even higher: TED = 3280 and ED is 2200 Wh/kg. A critical
discussion of the primary Li-nonaqueous batteries, as well as the
recent literature references, can be found in (22).

TABLE III.   Some Non-Aqueous Solvents for Candidate Storage-
             Battery Electrolytes

Solvent	Formula	M.P. °C	B.P. °C	D.C. 25°C	Max. K of 1M-LiClO₄ Soln. at 25°C
Propylene Carbonate PC	$CH_3-CH-CH_2$ with O, O, C=O ring	-49.2	241	64.4	0.0073
Butyro- lactone BL	CH_2-CH_2, CH_2 $C=O$, O ring	-42	206	39	0.015
Dimethyl Formamide DMF	$H-C-N$ with CH_3, CH_3, O	-61	153	36.6	0.021
Aceto- nitrile AN	$CH_3-C\equiv N$	-45.7	81	36	0.029
Methyl formate MF	$H-C$ with O, OCH_3	-99	32	8.5	0.032
Thionyl- chloride TC	$O=S$ with Cl, Cl	-104	77	9	0.020 (with 2M-LiAlCl₄)
. . . .					
Water	H, O, H	0	100	78	0.09
H_2SO_4 aq.					0.82 (25% soln.)
KOH aq.					0.60 (20% soln.)
Molten Salts: KCl; KOH; NaOH; KOH+NaOH eut.					0.75-2.5 (50°C above M.P.)

The swelling of the cathode $(CF_x)_n$ influences the discharge through the formation of a LiF precipitate (23). A film of LiCl is formed on the Li anode by its reaction with the depolarizer $SOCl_2$; this causes the voltage delay during discharge (24). Traces of water are thought to cause the formation of a film on the anode in the Li/dimethyl sulfite system (25). By 1973 there were advertisements that primary Li-organic (Li/SO_2) cells were commercially available (26), and indeed in 1978 they are available from a number of manufacturers. Even a D-size Li-CuS cell with a mixed nonaqueous solvent is superior to the aqueous Leclanché cell of the same size, according to the authors of (27).

A long-life, wide-temperature-range organic electrolyte cell, Li/PC-THF + LiAsF6 /AgCl, based on propylene carbonate and tetrahydrofuran was proposed in (28). The performance characteristics of $Li-SO_2$ and $Li-CF_x$ batteries were also discussed (29). Both showed excellent performance at low temperatures. However, some unsafe features associated with leakage of SO2 and/or internal shorting became apparent, and further efforts are needed before the Li-organic electrolyte battery can be considered fit for general use (29).

Other primaries have been proposed (30,31). Of all these, the most advanced is the Li-SO2 cell. The discharge characteristics of the Li-SO2 battery system are discussed in (32a); comparison was made with the other systems (32b). With a practical ED of 290 Wh/kg at room temperature, it can operate down to -53°C with ED = 55 Wh/kg. The SO2 is dissolved in acetonitrile + propylene carbonate in this case.

Butyrolactone is the preferred solvent of Japanese workers for primary batteries with carbon fluoride cathodes, although the K of the electrolyte at 45° shows a maximum of only 9×10^{-1} ohm cm^{-1}. The preparation of $(CF_x)_n$ and the ED dependence on the value of x in $(CF_x)_n$ are reported (33,34).

Published work on other metals (35), effects of moisture (36), unusual salt additives (37) and mixed electrolytes (38,39), round out the survey.

From a review of the recent Russian electrochemical literature it can be concluded that intensive research on nonaqueous batteries is carried out in the U.S.S.R. Although no results of the performance of such batteries are published, papers on properties of Li solutions in solvents used in Li cells, and on the behaviour of metallic Li in nonaqueous solutions, which have appeared recently in Russian literature, can be inferred to be the tip of an iceberg of research in that field. For example, the electrodeposition of Li from dimethylformamide solutions of its salts (40), or the behaviour of Li in aprotic solutions (41a) and the solubility and conductivity of its salts in these solutions (41b) emanate from an unnamed research institute in Moscow.

Secondary Battery Research. The fact that an alkali metal
can be electrodeposited from a nonaqueous solution of its salts
at ambient temperatures (3,4) suggests the possibility of
designing a SB based on that system with an alkali metal as
anode. If the metal can be regenerated by charging, it will work
again as anode. Many attempts have been made to bring about this
possibility.

The couple Li/CuSO$_4$, with ED = 1060 Wh/kg and emf of 3.41V,
has been studied with the aim of designing a SB with electrolyte
based on the sulfates and perchlorates of tetraammonium complexes
in PC and methyl sulfate. However, Li reduces the solvent (42),
and other more stable solutions are needed.

Attempts were made to utilize transition-metal sulfides as
cathodes in SBs. These sulfides, and especially those of Ti, are
electrochemically active and reversible in cells such as that
based on Li and TiS$_2$ in a LiAlCl$_4$-propylene carbonate
electrolyte. More than 80 cycles were performed with TiS$_2$, it
was reported (43a). Eleven hundred shallow (4%) cycles were
reported later (43b).

The chemistry of NbSe$_3$ and TiS$_2$ and their behaviour as
cathodes in reversible cells with Li and non-aqueous electrolytes
were studied at the Bell Laboratories. The reversibility of the
MX$_3$ cathode system is explained by the formation of intercalation
compounds between Li and the sulfide (44a). For example, in the
cell Li/LiClO$_4$ in PC/MS$_3$, where M = Nb or Ti, Li$_3$MS$_3$ is formed
and the cathodic reaction was found to be completely reversible
(44b). Scanning electron-microscopic studies of the cyclic
behaviour of this cell were made with NbSe$_3$, NbS$_3$, TaSe$_3$ and TiS$_3$.
Although all these systems show rechargeability, cell failures
are related to Li-anode morphology changes, accompanied by the
formation of a relatively non-adherent deposit of Li (45a). The
discussion in reference (45a) is of substantial interest, since
workers in the same field from other laboratories pointed out
causes of delay in the achievement of practical results with
Li-nonaqueous SBs.

The first crucial problem seems to be unfavorable morphology
of the Li deposits, since Li is intrinsically reactive in all the
solvents that are of interest for Li-organic electrolyte SBs.
This reactivity does not appear with bars or foils of metallic
Li, but only with a cycled Li electrode. If rechargeable Li SBs
are to be practical, a means to control the morphology of Li must
be found (45b).

The second critical problem is the chemical instability of
Li which deposits during the cycling of secondary cells.
Electrodeposited Li has such a high surface area that it is not
stable in many solutions in which flat Li foil is stable (45c).
However, the pessimistic opinions of (45b and c) have not
inhibited the authors of (45a) from claiming a patent for a
nonaqueous battery using chalcogenide electrodes, the specific
structure of which is the main feature of the SB (46).

The reversibility problem of transition metal chalcogenides in Li nonaqueous batteries has attracted the attention of other researchers who have studied the structure of these compounds and the complexes they form with Li (47). The relations between the structure of the ternary phases formed at the cathode of a Li-transition metal sulfide, or oxide, and the reversibility of the cell are discussed in (48). The reversibility of the discharge reaction is maximized when no chemical bonds are broken during discharge (intercalation reaction) and it is minimized when all the chemical bonds are broken, such as in the case of CuS or CuF_2. When only some, but not all, of the chemical bonds are broken and the structure is distorted as in TiS_3 and V_2O_5, partial rechargeability is found. Thus cells of the type $Li/LiClO_4$-organic solvent/V_2O_5 have been studied with various solvents. PC supports the greatest number of cycles. After about 400 cycles failure occurred, due to the swelling of the Li electrode as a result of its low density after being deposited at the negative electrode (49). A reversible cell with a dissolved cathode of Li_2Sn and electrolyte based on dimethyl sulfoxide and tetrahydrofuran as a mixed solvent, Li/DMSO or THF + $LiClO_4/Li_2S_n$, was studied in (50). A nonaqueous $Li-Br_2$ secondary cell: $Li/LiBr-PC/Br_2$ was studied in (51). With an OCV of 3.82V it has an ED of 704 Wh/kg. More than 1700 cycles were achieved. The Br_2 electrode was still completely serviceable but the Li electrode had partly disintegrated.

The authors of (49) were granted a U.S.A. patent on a rechargeable battery described above. The electrodes are enclosed in special wrappings to prevent the passage of large Li particles (52). A patent was granted to the authors of (45a), who claim that the addition of certain dopants to their battery will improve its performance, especially the cycling (53).

Although most of the publications on nonaqueous SBs are concerned with Li anodes, there are several articles and patents on secondary cells with other light metals. Thus a cell Na/PC + NaI/Na_xTiS_2 was described in (54), a German patent was taken out for a nonaqueous rechargeable Na-halogen battery (55), and a number of Japanese patents have been granted for nonaqueous batteries with Li, Mg or Al anodes (56). Nothing is said about cycle life, but good capacity is indicated.

<u>Prospects for SBs Based on "Nonaqueous" Electrolytes</u>. Despite the availability of several articles and patents on nonaqueous secondary batteries (42-56), there is little hard evidence that these batteries are approaching the commercial stage. However, the fact known long ago that light metals of high energy density can be electrodeposited from their nonaqueous solutions at ambient temperatures (3,4,21), and the successful completion of several primary nonaqueous batteries to the commercial stage, has raised hopes of the possible construction of secondary HED batteries based on the same solutions. So far,

considerable difficulties and obstacles have been met during the
work on rechargeable batteries. One obstacle, which is common to
both primary and secondary batteries, is the low conductivity of
nonaqueous solutions, which in the best cases is less than 0.025
$ohm^{-1}cm^{-1}$. However, this difficulty can be circumvented by using
electrodes of large surfaces with close distances between them.
The second difficulty, which is specific for the secondary
battery, limiting its cycle life, is the state of the anode after
recharging. The brittle, crumbling, regenerated anode,
impregnated with a gum-like substance, will have electrochemical
properties far different from the metallic bar or foil of the
uncycled, fresh anode. Pointing to the changing morphology of
the Li anode during cycling, the author of a 1976 review stated
(19) that on the whole the practical secondary Li-electrode in
organic electrolytes will probably remain an unachieved ambition,
at least for the near future, although several types of primary
Li-batteries are already commercially available. Perhaps the
answer will be found with new solvents, such as those proposed by
Caiola et al. (57).
        It is of interest to note that, despite these and some other
pessimistic appraisals of the Li-nonaqueous SBs for the future,
at the XIth Mendeleev Congress in Alma-Ata (Kazakhstan) in the
presence of many hundreds of chemists (September 22-27, 1975)
A.N. Frumkin, just a few months before his death, recalled that
among the most optimistic opportunities in applied
electrochemistry are the creation of fuel cells for continuous
power and of high-energy-density storage batteries based on
aprotic solvents and alkali metals (58). And there are many
European and North American enthusiasts who agree, as the
references attest.
        There are at least two ways to overcome the difficulties
connected with the reversibility of SBs based on nonaqueous
electrolytes, quite apart from the usual trick of supporting the
electrode materials:
(a) Investigate a number of other solvents and their mixtures in
    the selection of electrolytes for SBs.
(b) Create conditions under which the changes in the active
    metal's morphology during recharge would not influence
    the cycle life of the cell.
        The ten or so organic solvents and the 4 or 5 inorganic ones
(Table III) tested in experiments with nonaqueous SBs represent
only a small fraction of the liquid compounds which may prove to
be useful as a basis for electrolytes in these batteries. The
problem of nonaqueous solvents for electrochemical use is
discussed in (59), where nine groups of organic and inorganic
liquids -- nitriles, amides, amines and so on -- were considered.
They include also such compounds as alcohols and acids which have
not been tested for SBs, being not aprotic. In any event such
compounds, especially in mixtures with aprotic ones, should be
examined. Special attention must be given to inorganic compounds

which can play a double role, both as solvent for the conducting species and as oxidizer, i.e. an active cathode material. Hydrazine (60) and acetamide (61) could be considered as candidate solvents, but in formamide (62) lithium deposits become crumbly.

There are many solvents which have not yet been tried in cells with light metals intended to be reversible. Various systems of solvents (some of them containing even several percent of water) are often more suitable than their components in revealing values of properties beyond those of the components. Much more attention must also be given to the role of the complex formation and mass-transfer just prior to the electrodeposition of the active metal during recharge. A systematic and thorough kinetic analysis of the relationships amongst the various elementary steps of the metal redeposition process is badly needed -- otherwise we may search a long time without finding the probably quite narrow set of conditions under which morphological reversibility will occur over a large number of cycles.

Finally, the exclusive attention to Li as the anode in HED nonaqueous SBs at the expense of the other five light metals cannot really be justified. Although its reactions have higher theoretical EDs, the difference between the ED available with it and with Na, Ca, Mg and Al, when the weight of the whole cell with its accessories (electrolyte, conductors, separators and casing) is taken into account, is really quite small. These metals, and even Ti, may show such advantages in their electrochemical behaviour (especially in cycle life) which may counterbalance their comparatively lower TED's.

Molten Salt Electrolytes

The theoretical and experimental work on electrochemical properties of molten salts done up to the 1910's was summarized in the monographs of Richard Lorenz, who himself made important contributions in that field (63). Further developments in molten salt electrochemistry occurred in the 1930's in connection with the electrochemistry of light metals, Na, Mg and especially Al. Many books, papers and reviews were published in the English, German and Russian literature, from that time on (see, for example, (64,65)). The development of nuclear power in the 1950's and the 1960's re-stimulated experimental work on molten salts, which were suggested as coolants and heat-transferring media in atomic reactors. The papers from the US Bureau of Mines and from the Atomic Energy Commission are rich and numerous. Several books and reviews which have appeared recently contain chapters on electrochemistry of molten salts (66). A review of galvanic cells in molten salts with 351 references was published in 1968; most of the works cited there deal with the potentials of various electrodes in fused salt media (67). A review of molten salt batteries and fuel cells was published in 1971 (68).

266

CHEMISTRY FOR ENERGY

The development of the Molten Salt Information Center at
Rensselaer Polytechnical Institute, by G.J. Janz, has been a boon
to researchers in this field.

Use of Light Metals. Tables of the open circuit voltage
(OCV) and of theoretical EDs are presented in (68) for
hypothetical cells formed by light metals with halogens, oxygen
and sulfur at 500°C. We have condensed these into Table IV.

TABLE IV. OCV and TED (KWh/kg) of Cells at 500°C

	F		Cl		Br		I		O		S	
	V	ED	V	ED	V	ED	V	ED	V	ED	V	ED
Li	5.56	5.74	3.65	2.31	3.21	0.99	2.57	0.57	2.14	3.84	2.3	
Na	5.12	3.27	3.52	1.61	3.16	0.82	2.59	0.50	1.62	1.40	1.82	1.25
K	5.02	2.32	3.76	1.35	3.48	0.78	2.98	0.52	1.27	0.72		
Mg	5.01	4.31	2.68	3.57	2.26	0.66	1.61	0.34	2.68	3.56		
Ca	5.60	3.84	3.53	1.70	3.06	0.82	2.48	0.49	2.88	2.75	2.33	1.71
Al	3.87	3.71							2.48	3.91		

The voltages of Ca and its EDs are noteworthy: they are
higher than or close to those of Na. As to Mg and Al, they too
present values of V and ED close to those of Na. The ED for the
chloride, for example, of Mg is even higher than that of Li. And
as the V's and ED's of practical cells are lower than those of
the hypothetical ones because of the IR losses (for V) and the
weight of the whole cell (for ED), the practical figures for
these light metals are even closer. However, as is shown in
Table V, also based on (68), the cell voltages, the power
capacities, and the EDs of the molten salt experimental SBs are
much higher than those of the conventional ones.

Two earlier reviews were published on high temperature cells
and batteries based on molten salt and solid electrolytes. The
first one (69) describes the Li/Cl_2 cells, particularly the
$LiAl/LiCl-KCl/Cl_2$ cell with gaseous Cl_2. Li cells with
chalcogenides as cathode materials are mentioned, as well as some
details of construction. This review, and the 26 references
attached to it, reflects the state of the Li molten salt
batteries to the end of 1970 (69). The second review (70),
prepared two years later is more comprehensive. It discusses in
detail some theoretical problems, the thermodynamics and rate
processes in electrochemical cells, and presents tables and

TABLE V.   Comparison of Properties of Conventional and New
Types of SBs.   (See Ref. 68)

Battery	T °C	Cell Voltage	Power Capacity W/kg	ED Wh/kg	Cycle Life
Pb-acid	ambient	2.1-1.5	6-30	20-30	100-400
Ni-Fe	"	1.3-0.8	7-40	30-35	100-3000
Ni-Cd	"	1.3-0.8	7-45	36-40	100-2000
Ag-Zn	"	1.35-1.1	25-150	80-100	100-300
Zn-air	"	1.4-1.0	40-60	100-150	
Na-air	130	2.6-1.8	80-100	180-273	
Li-S	340	2.3-1.2	550-800	250-360	
Li-FeS	340	2.0-1.2	∿ 600	120-200	100-800
Na-Bi	350	0.8-0.4	80	40	
Na-S	300	2.1-1.2	200-350	180-330	50-2000
Li-Cl$_2$ (G.M.)	650	3.5-3.0	200-400	300-400	
Li-Cl$_2$ (Sohio)	450	3.4-1.0	330-440	100-180	225
Li-Te	470	1.8-1.0	550-800	185-265	
Li-Se	375	2.2-1.2	600-1000	220-330	

graphs characterizing the high-temperature secondary cells. Five
such cells, with molten salt electrolytes and Li anodes, are
presented in Table VII of that review, and the cycle-life and the
state of development, along with the literature cited, are
indicated.

The operating temperature of such cells is from 340° to 650°C
and the cycle-life can vary widely: 100 cycles for the Li-Al/TeCl$_4$
(C) cell, to >220 for the cell Li-Al/Cl(C) for example. These two
cells are in the laboratory development state, being evaluated as
cells and batteries. Three others, based on S, Se and Cl are
still in the early experimental stage, as single cells. The
number of charge-discharge cycles of the secondary cell depends,
for the same system, upon the current density and on the depth of
discharge. Generally, molten salt cells withstand deep discharge
cycles better than aqueous or non-aqueous cells, it seems.

The problems connected with the TeCl$_4$(C) and Cl(C) cells are
low capacity and parasitic currents, and of the S, Se and Cl$_2$
cells corrosion of the seals and insulators. All these cells
were (in 1973) in the early stages of development. The cells
have ED of about 200 Wh/kg (70).

We now consider some publications on molten salt SBs which
have appeared since 1973.

Li and Li-Alloy SBs. Of all the light metals, Li shows the
least tendency to dissolve rapidly and easily in molten salts, a
favorable fact for Li cells. Li dissolves in LiCl at 640° up to
0.2-0.5 mole %, whereas the solubility of Na in NaCl at 811° is
2.8 mole % and of Ca in CaCl$_2$ at 1000° is 5.4 mole % (70a).

The two Li-Cl$_2$ cells shown in Table V, which operate at 650°
and 450° respectively, have required substantial improvements
directed to their possible practical use. The first one, studied
in the G.M. laboratories, has seen its operating temperature
lowered from 650° to 425° by the utilization of the ternary salt
mixture LiF-LiCl-KCl. The cell Li(liq.)/LiF + LiCl + KCl/Cl(C)
(porous graphite), which gave 232 cycles and about 350 Wh/kg ED,
was recommended as one of the most promising high-temperature
SBs for vehicular propulsion (71).

However, more publications have appeared on Li SBs with
sulfur or sulfide cathodes. For example, a cell Li/LiCl-KCl
eutec./S-As, in which arsenic was added to retain the sulfur
inside a molybdenum-mesh support, was cycled (72). The same
electrode materials, with LiF-LiCl-KCl as electrolyte, were studied
in the Argonne Labs for application to electric automobiles. The
ED is >200 Wh/kg and specific power >200 W/kg, when operated at
380 to 425° (73).

Pure S is unsuitable because the operating temperature of
these cells is so high (about 400°) that S dissolves in the
LiCl-KCl electrolyte and reacts with Li at the anode, leading to
self-discharge. However, in metal-sulfide cells these problems
have been circumvented. Attempts have been made to replace S by

sulfides of Fe, Ni, Co and Cu; several disclosures have been
issued on this topic (74,75). In Table VI the OCV and the
theoretical ED for such SBs, are given, as well as the formulas
of the cathode material before and after discharge. Up to 300
cycles (for Cu-S up to 912) have been achieved (76). At the
temperatures of the electrolytes (350°-400°), Li is liquid and
highly corrosive.

TABLE VI. Li-MS$_n$ SBs. Theoretical ED's, OCV, and the Composition of the Cathodes Before and After Discharge

Sulfides	Fe-S	Cr-S	Ni-S	Cu-S
Probable composition of the cathode before the discharge	$FeS_{1.5}$	CrS	Ni_3S_2	Cu_2S
Discharged Cathode	Fe	Cr	Ni	Cu
Theoretical ED Wh/kg (Li-MS$_n$)	1153 (968 for FeS_2)	693	678	518
OCV	1.62	1.27	1.70	1.68

Therefore many attempts have been made to use Li alloys which
are solid at high temperatures. The compound LiAl melts at 718°,
i.e. it is solid at the operating temperature of the SB, which is
usually below 450°. Experiments have been carried out on the
anodic dissolution of this alloy: it shows a voltage between
-1.95 to -2.0V at a Li content between 14.7 and 48 atom % (77);
and SB cells of more than 120 Wh/kg ED, intended for off-peak
energy storage and for car propulsion, are under development.
The failure mechanism and capacity loss, due to the disintegration
of the Al-Li alloy during cycling, were studied; and it has been
suggested that pyrometallurgically prepared alloys might show
better results (78), but this awaits demonstration. Further,
X-ray radiography was used to study the failure mechanism and
capacity loss due to disintegration of the Al-Li alloy during
cycling of the cell Li-Al/KCl-LiCl/Te (79).

Li-Al anodes have been combined in cells with Cl_2 in the
Sohio Carb-Tek battery, operating with a molten salt electrolyte
in the range of 400°-500°C. A porous carbon cathode and a BN
separator were used. Addition of $TeCl_2$ to the positive electrode
increased the capacity in the 3.25-2.5V range. Although the
battery presented many problems associated with the materials of
the electrode, the casing and the seal, corrosion by Cl_2 being

especially troublesome, it was found that, in an O_2-free electrolyte, vanadium may serve as a stable current carrier (80). Li-Al and Li-Si alloys were compared in (81). The structure and anodic discharge behaviour of Li-B alloys in the LiCl-KCl eutectic melt are discussed in detail in recent articles. In one study, from the shape of the discharge curves the existence of a compound Li_2B in the Li-B system was surmised. The experiments have shown that Li-B alloys promise good results (82).

The data in some very recent publications indicate that long life may very well be achievable in some Li-molten salt systems. Particularly encouraging is the work reported in Proc. IECEC in 1977 by the teams from Argonne National Laboratory, General Motors Research Laboratory and Atomics International, under the stimulus of the U.S. Government's Energy Research and Development Administration. In the earlier symposium, at Argonne National Laboratories in 1976, more detailed reviews of the several approaches being made by these groups were given. At this rate of effort and progress the capability for industrialization of molten salt HED battery systems indeed seems feasible by 1983, as the U.S. Department of Energy (formerly ERDA) apparently fully intends.

The Other Five Candidates. All the molten salt SBs reviewed above have either a Li anode or a lithium alloy, one in which Li prevails quantitatively. As to the other 5 light metals they are seldom mentioned in the literature as candidates for anodes in these SBs, except Al. In (82) it is stated that molten salt batteries with Ca or Mg anodes yield only a small proportion of their theoretical energy because (a) Ca anodes react chemically with the electrolyte, and (b) both Ca and Mg anodes are passivated at high current drains, becoming coated with resistive films of solid salts. In a melt containing Li salts, Ca replaces Li ions by the displacement reaction $Ca + 2LiCl = CaCl_2 + 2Li$.

The Li metal so produced will form a liquid alloy with Ca (m.p. 231°) in thermal batteries (83a), eg. $Ca/LiCl-KCl-CaCrO_4/Fe$, to which the heat needed to melt the electrolyte is provided by $Fe + KClO_3$ pellets (83b). More recently a rechargeable battery based on calcium was described: $Ca-Al/CaCl_2 + NaCl/FeS$. Although its OCV is only about 1.86V, the cell was shown to be capable of sustaining some tens of cycles, and show promise (84). (See the several other papers along this line in the same Proceedings of the excellent Symposium at Argonne.)

In a U.S.S.R. patent, Na anodes in molten salts (NaOH + NaBr) were disclosed as a source of electric power at high temperature, but no cycling data were presented (85a). A secondary battery operating at 150° with high TED and achieved energy output is described in a German patent, based on the system $Na/NaAlCl_4/C$, with a beta-Al_2O_3 separator (85b).

There are several publications on molten salt SBs with Al anodes. $AlBr_3$, the Al halide with the lowest m.p. (97.5°), mixed

with one or more alkali metal halides, was offered as an
electrolyte for the battery $Al/AlBr_3$ + MX/C, operating in the
temperature range 50–250°, but nothing is said in the patent
abstract about cycling (86).

Table VII gives the m.p. of other alumohalides and their
mixed systems. For example, low-melting electrolytes based on
$AlCl_3 \cdot MCl$ chloraluminates, where M is Li, Na, K, have been
considered (87), and cells with Al anode and various cathodes,
both inorganic and organic, were tested. The sulfur cathode seems
to be the most suitable, although complex chlorides, fluorides
and sulfides show possibilities. An experimental Al/S cell is
described in detail in (88). The reaction: $2Al + 3S = Al_2S_3$
provides a TED of 1275 Wh/kg at 200°. It is viewed only as a
primary battery, however at the present time (88).

The feasibility of the use of titanium in rechargeable SBs
has been the principal topic of a recent report prepared by the
authors. The situation can be summarized as follows. Although
Ti can be dissolved in molten salts and deposited from these
solutions by electrochemical reduction (89), the present authors
have not found any publications in which Ti is offered as the
rechargeable negative plate in molten salt SBs. Perhaps no
serious attempts have been made to investigate the possibility of
developing cells with Ti as negative plate, either in nonaqueous
or molten salt electrolytes; at least, nothing of such attempts
has been published. There are however, several publications
discussing the use of Ti compounds in the cathodic material, or
Ti alloys in supports, casing and conductors. In Table I of
(90a), titanium is listed as an anode among other candidate
metal-oxygen systems. Its voltage and EDs in such systems are
among the highest, and the cost per kWh the lowest, but no
indications of its use were claimed in (90a). Our detailed
review of possibilities is given in (90b), and the reaction with
$Ni_{12}P_5$ in molten $NaPO_3$ is reported in (91).

In Tables VII, VIII and IX, which are based on the data given
in (92–94), are listed some other candidate molten salts,
inorganic and organic, worthy of consideration as electrolytes.
It will be noted that there are many low-melting molten salts
whose potential to form complexes with the light-metal ions is
high, and which have not been examined experimentally as to the
throwing power for electrodeposition (recharge) of the active
metal. Further from his computerized listing of properties of
molten salts, Janz has tabulated an even more impressive list of
possibilities (93), some of which have acceptable conductivity and
thermal stability as well. Further a priori selection of suitable
candidates might perhaps be based on knowledge of the nature and
extent of complexing with the active metal ions, were such information available.

TABLE VII.   Alumohalides and Their Systems [92]

A. Individual Alumohalides

Pure Compounds	M.P. °C	Boiling P. °C
AlF_3	445	1272
$AlCl_3$	192.4	187
$AlBr_3$	97.5	253
AlI_3	191	380

B. Systems

A	B	C	M.P., °C	Mole % B	C
LiCl	– $AlCl_3$		105	58	
NaCl	– $AlCl_3$		110	61	
KCl	– $AlCl_3$		128	67	
NaCl	– $AlCl_3$		150	50	
LiBr	– $AlBr_3$		105	(75)	
LiBr	– $AlBr_3$		195	50	
NaBr	– $AlBr_3$		95	76	
KBr	– $AlBr_3$		191	50	
NaI	– AlI_3		123	70	
KI	– AlI_3		105	67	
$NaCl \cdot AlCl_3$	– $NaI \cdot AlI_3$		80		
$AlCl_3$	– NaI	– AlI_3	65		
NaCl	– KCl	– $AlCl_3$	88	16.5	63.5

TABLE VIII. Low-Melting Inorganic Compounds and Systems [93]

A. Compounds

Compound	M.P., °C	Compound	M.P. °C
$TeCl_2$	175	BiI_3	43
$MoCl_3$	194	SbI_3	170
$InCl$	225		
$InCl_2$	235	NH_4CNS	87.7
		$KCNS$	171
$LiCLO_3$	127.6		
$LiClO_4$	236	NH_4NO_3	169
		$NaFeCl_4$	163
$SbBr_3$	96.6	$KFeCl_4$	250
$GaBr_3$	124.5	$NaNH_2$	208
$SnBr_2$	232		

B. Systems

A		B		C	M.P. °C	Mole % B	Mole % C
$LiNO_3$	–	NH_4NO_3			97	75	
$TiCl$	–	$CuCl$			122	60	
$KCNS$	–	$NaCNS$			123.5	70	
$LiNO_3$	–	KNO_3			125	44	
KCl	–	$SnCl_2$			180	62	
$NaOH$	–	KOH			187	40	
$LiNO_3$	–	NH_4NO_3	–	NH_4Cl	86	66.7	7.5
$NaClO_3$	–	$LiClO_3$	–	$LiNO_3$	92.6	(50)	(40)
$LiNO_3$	–	KNO_3	–	$NaNO_3$	120	53	17

TABLE IX.   Organic Salts and Systems [93]

A. Compounds   M.P. °C

    TlCOOH   100.8
    KCOOH   167.5
    TiCOOCH$_3$ 130

B. Systems

		Mole %		
		B	C	D
KCNS-KHCO$_2$	78.5	54		
Ca(HCO$_2$)$_2$-2TlHCO$_2$	94.2	91.2		
Sr(HCO$_2$)$_2$-2TlHCO$_2$	96.8	94.9		
KCNS-KHCO$_2$-KNO$_3$	55.5	48.3	16.8	
(Cs,K,Li)CH$_3$CO$_2$	73	27	20	
KBr-KCNS-KHCO$_2$	75	43	55	
KCl-KCNS-KHCO$_2$	76	43.5	55	
K,Li-HCO$_2$-CNS	53	50	45	5

Prospects for SBs Based on Molten Salt Electrolytes.   The main advantage of molten salt electrolytes for HED SBs is the fact that each of the selected six most active light metals -- Li, Na, Ca, Mg, Al and Ti -- can work reversibly in molten, mixed electrolytes, i.e. they can be.anodically oxidized during discharge, and cathodically reduced by recharging.   Further, some materials like sulfides have been found to be reversible as cathodes.   Another advantage of the molten salt electrolytes is high conductivity -- about 100 times that of the same salt in non-aqueous solvents, and 5-10 times higher than their aqueous solutions.   There are two intrinsic shortcomings which result from the high operating temperature: (a) high rate of corrosive side reactions, which can be circumvented only by judicious choice of often expensive materials, and (b) the inevitable heat management required.

The Li-Al/FeS$_x$ system is being heavily studied currently, and is being built into a vehicle-demonstration unit.   It is one HED system which does show excellent promise as a long-lived storage battery.   This important breakthrough suggests that there may be other systems even more suitable.   Thus researchers have been exploring lower melting molten salt electrolytes, the mixed

halides of aluminum, eg. (85-88), and there are many other molten
salt electrolytes which are liquid in the temperature range
60°-125°C (Tables VIII and IX). It will be noted that amongst the
low-melting systems there are many which contain the cations Li^+,
Na^+, K^+ and Al^{+++} and the anions Cl^-, Br^- and NO_3^-. Tables VII to
IX cover only a fraction of all the low melting systems: studies
on others have been published in the literature.

In Table X we give a condensed list of electrolytes which
need to be examined experimentally as potentially capable of
supporting reversible electrochemical processes at anode and
cathode. Still many more possible combinations have not yet been
investigated, such as those containing Ca^{++}, Mg^{++} and Ti^{++}
cations, as well as HSO_4^-, HSO_3^-, HCO_3^-, and some other anions.
The compatibility of each light metal cation with any low-melting
molten salt electrolyte must be examined experimentally. It can
be anticipated that at least some such combinations will have
useable properties.

TABLE X.   Some Molten-Salt Candidates as SB Electrolytes
(See Table I.A.4 of (93))

Electrolyte		M.P. °C	Status/Use
cryolite	$NaF \cdot AlF_3$	886	commercial winning of Al
"salt"	NaCl	800	commercial electrodep. of Ti
flinak	LiF-NaF-KF eut.	454	electrodep. of V, Mo, W, eg.
cllik	LiCl-KCl eut.	357	basis of Li/FeS storage bty
nitrates	$LiNO_3$-KNO_3 eut.	191	unknown as SB electrolyte
caustic	NaOH-KOH	187	basis of H_2/O_2 fuel cell
formates	$Ca,Ti(HCO_2)_2$ eut.	94	unreported
fcn's	$KCNS$-$KHCO_2$	78	probably untried

The availability of low-melting salt combinations opens up
potentialities for all the light metals. As we have seen above,
almost all the research work on HED SBs with molten salt
electrolytes is carried out with Li as an anode and with
chlorides as electrolytes. Other light metal systems should be
investigated too, especially in view of the much greater natural
abundance of some of these other metals.

Solid State Electrolytes

Like the Li/FeS$_x$ system, which is presently the most advanced
rechargeable battery system based on a molten salt electrolyte,
the Na/S system is presently the most advanced rechargeable
battery system based on a solid electrolyte (beta-alumina). It
operates at about 300°C.

Appreciable ionic conductance of several solids has been
known since the turn of the century when the uniqueness of AgI
was uncovered. Since then, doped iodides, mixed inorganic salts
with super-lattices, and even ionically-conducting organic
polymers, have been discovered or synthesized. Serious attempts
to use these materials in storage batteries are quite recent,
however; the only really successful application in a
rechargeable battery so far is that of beta-alumina. Solid
electrolytes can be made thin, generally, as can cells made from
them. Some of the materials have high conductivities, even at
temperatures close to ambient, the silver ion conductors, for
example. The search is for one through which one of the light
alkali metal cations has high conductivity at normal temperature
or at an elevated temperature easily maintained operationally:
one which can be made thin, ideally flexible or not friable,
whose chemical composition is stable, and with which solid-state
contacts can be maintained by anodic and cathodic materials
during repeated cycling.

A very instructive perspective concerning cell design
concepts from which rechargeable cells based on solid electrolytes
can be viewed has been offered by Armand, Univ. de Grenoble.
Modified somewhat, it is given in Figure 2. Note how properties
of the electrolyte can often dictate the design of a storage
battery.

Complex Compounds of AgI. A historical review of the
conductance studies of these compounds and the emf's of cells
based on these solid electrolytes, was published in 1969 (95).
Another review, which discusses the conductance and the structure
of various solid electrolytes (AgI and its derivatives, Ag$_2$S,
Li$_2$SO$_4$, Na$_2$O·11Al$_2$O$_3$) was published in 1972 (96). There have been
several papers published since, especially by Italian and
Japanese workers.

As a rule, one solid phase of these electrolytes is a fairly
good conductor, its open structure allowing the cation (Ag$^+$, Na$^+$,
Li$^+$) to move fairly freely inside the lattice. In the case of
AgI the conductivity, K, increases by two orders of magnitude at
147°C, as the beta-modification is transformed into a more open
alpha-structure. At ambient temperature the conductivity of AgI
itself is very small, but it is much higher in some complexes,
such as RbAg$_4$I$_5$, in which K = 0.2 ohm^{-1} cm^{-1} at 20°C, for
example. Since the discovery of the high ionic conductivity of
RbAg$_4$I$_5$ (97a) much work has been done on solid electrolytes.

Takahashi and coworkers have studied the conductance and the crystal structure of AgI complexes, as well as their behavior in electrochemical cells (97b).

Values of K as high as 6×10^{-2} ohm^{-1} cm^{-1} have been found in combinations of AgI with substituted ammonium iodides (98). These studies were extended and the results used in the development of primary solid-state batteries, for operation at ambient temperatures, based on cells such as Li/LiI-CaI/AgI(C) (99a), and with Ag_2CrO_4, Ag_3PO_4 and the superlattices formed with AgI (99b).

The properties of the AgI-based solid electrolytes have been studied also in the U.S.S.R., and several publications have appeared. The research was directed towards the use of solid electrolytes in batteries and in information devices (100a). A thin-film galvanic battery with solid $RbAg_4I_5$, and based on Ag + Te was studied (100b), and several publications describe the construction and properties of that cell (101). The method of synthesizing the compound $RbAg_4I_5$ by precipitation in acetone (102) is also described. A recent authoritative review on these and other compounds is now available (103). See especially Table IV in that review.

Although primary batteries of various types based on AgI complexes are at present available commercially, no substantial success has been achieved with rechargeable SBs with Ag compounds conducting at ambient temperature (although Ag/complex/Ag coulometers capable of being cycled many thousands of times are readily available). After a decade of studies, cells based on AgI complexes have not yet been made which are rechargeable to any extent: the complex tends to break down into islands of the γ-AgI, which is very poorly conductive at room temperature.

The Beta-Alumina SB. The beta-alumina cell from the very beginning was considered to be a SB. The first detailed publications on it appeared in 1967 (104), and since that time scores of open publications and patents and several reviews on this SB have been issued. In the reviews cited above on high temperature cells (69,70) substantial portions are devoted to this SB; the structure and properties of its solid electrolyte, $Na_2O \cdot 11Al_2O_3$, are discussed in (96). The cell has a liquid Na anode and a liquid S cathode. The reaction product is the sodium polysulfide, of varying composition and physical properties. The good performance of the cell is based on the fact that the electrolyte is easily penetrable by Na^+ at reaction temperatures of about 300°C. During the charging process sodium ions are carried back through the beta-alumina electrolyte to the anode where they are reconverted to Na metal. In practice polysulfides of Na are formed at the cathode. The high operating temperature of the Na/S battery is necessary not only because of the high m.p. of the Na polysulfides formed at the cathode, but also because the conductivity of the beta-alumina is better the higher the

temperature. Attempts have been made to decrease the operating
temperature by incorporating MgO or FeO into the electrolyte
(105). Many patents have been granted to workers who claim that
the SB can be improved by other modifications (106). Results of
extensive research work world-wide are available (107).

The history of the beta-alumina SB development and the state
of the problem in 1974 are presented in Ref. (107a), which
includes references to the most important literature up to that
time. The processes at the cathode are complicated. The Na-S
phase diagram has been investigated: two eutectics (at 235° and
257°) were found and the region of the two liquid layers was
bracketed at 78 - 100% S (103). The paper seems to be one of the
most comprehensive and substantial articles yet to be issued on
that topic. It is believed that the Na-S SB has a potential
energy saving of a factor of nine over the lead-acid battery and
could give a range of more than 100 miles to electric cars.
Electric buses with a 40-mile operational range have already been
built (108). Although the time-table has slipped somewhat as
materials problems have been uncovered and resolved, it is very
probable that the world-wide effort on the Na-S SB, may yet be
crowned with successful application.

The shortcomings of solid electrolytes -- the sensitivity of
the beta-alumina batteries to vibration and shock, and the low
conductivity, the high cost and low cyclability of the AgI based
cells -- have stimulated the search for new types of solid
electrolytes. In the 1972 review (109a) of the advances in Li/
solid-state batteries, most of the cells discussed contained Ag
and Li/LiI/I_2-benzidine, are also discussed (109a). The
iodine-benzidine cell is described in some detail in (109b), and
the iodine cell in (109c). The properties of the high temperature
phase of Li_2SO_4 as a solid electrolyte are considered in (110).
So far, such cells have found only limited practical applications,
and these are primary batteries with very small currents and
long shelf life. Little success has been realized from attempts
to make mini-SBs based on solid electrolytes.

A number of publications have appeared recently on
super-lattice complexes which have enhanced conductivity, eg.
"nazirpsio": $Na_3PO_4 \cdot 2ZrO_2 \cdot 2SiO_2$ whose conductivity at room
temperature is of the same order as that of an aqueous salt
solution. Most of the super-lattices are unstable thermodynamically, and can be expected to collapse under chemical
attack by the anodic and cathodic reactants. However, there may
exist some thermodynamically stable structures, and the search
should concentrate on the complicated phase-diagram studies of
selected quaternarys.

In the Institute of New Chemical Problems in Moscow, the
resistance at the boundaries between Na and $Na_2O \cdot MgO \cdot 10Al_2O_3$ was
studied (111a). A patent for a solid electrolyte containing NaF
and Al_2O_3 was granted (111b). These solid electrolytes were

synthesized by a gas-flame method. Those with the composition $Na_2O \cdot MgO \cdot nAl_2O_3$ (n = 5 to 12) possess ionic conduction of about 0.1 ohm^{-1} cm^{-1} at ambient temperature; this value decreases when Na_2O is partially replaced by Li_2O or MgO (112).

A small amount of Sb_2O_3, added to the sulfide cathode of the solid Al_2O_3 battery, decreases the melting point of the sulfide, while not impairing the reactions of the cell (113). Other compounds have been briefly described (114).

Solid Polymer Electrolytes. Some solid polymer electrolytes are listed in Table XI. The most celebrated, "nafion", a H^+ conductor, was early recognized as a suitable electrolyte for the H_2/O_2 fuel cell, and more recently for hydrogen/oxygen or hydrogen/halogen storage battery systems. Attempts are being made by the US Department of Energy, through Brookhaven National Laboratory (115), to learn how to handle the mass-transfer problems in the "electrolyzer/fuel cell storage-battery" concept. Estimated achievable energy densities are respectable, above 200 Wh/kg. Although the turnaround efficiency expected for H_2/O_2 is less than 50%, the emergence of nafion-like materials permits the serious design and development of the hydrogen/halogen electrical-storage-battery systems, first proposed in 1964 (116), to be considered. Load-levelling SBs can be envisaged, in which the efficiency may exceed 70% (115).

By contrast, NaI-doped polyethylene oxide membranes have permitted experimental research on tiny rechargeable Na/I_2 batteries to be initiated (Figure 2). Chemical stability of the electrolyte, and the integrity of the mechanical contacts at the current collector/electrolyte interfaces, during repetitive cycling, must be improved.

The Prospects For Solid Electrolyte SBs. For reasons discussed above, the AgI-based cells, being useful for some special types of primary batteries, are not very promising for secondary ones. The beta-alumina cells, on the contrary, have already been developed to the pilot-plant stage and their prospects are fairly good to become commercialized. They are the most advanced among the candidate batteries for traction. The high operating temperature could be lowered if a solid electrolyte with K values of about 0.1-0.3 ohm^{-1} cm^{-1} at lower temperatures could be found. Since the metal ion (Na^+, Li^+) moves through the interstitial vacancies, the structure of crystals with such vacancies is the object of study. Such compounds might be found among silicates; thus even at present some Li-containing silicates, Li_4SiO_4, for example, are known for which K = 9.0 x 10^{-4} at 400°C. $LiAlSiO_4$ has a K-value of 5.6 x 10^{-5} (114a). Among the great variety of silicates, perhaps other, better-conducting complexes can be found. Dehydrated zeolites of special structure might be found which can act as a ion-exchanger electrolyte, containing ions of any light metal in the series

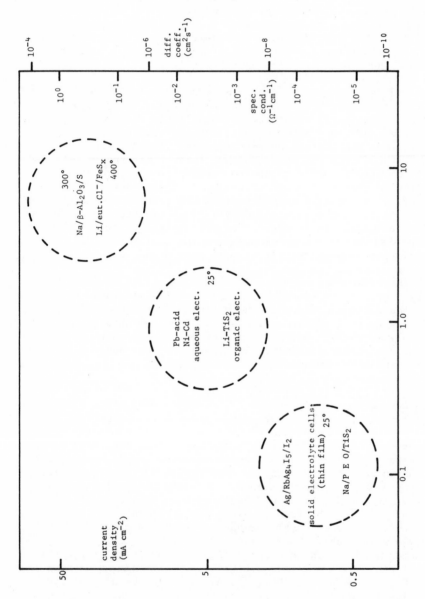

Figure 2. Cell-design concepts for new electrolytes (after Armand)

TABLE XI.   Some Solid-Polymers as Candidate SB Electrolytes

---

perfluoro carboxylic acid (asahi)

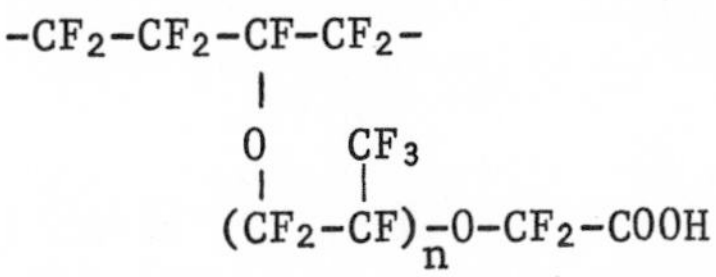

$$-CF_2-CF_2-CF-CF_2-$$
$$|$$
$$\quad O \quad\ CF_3$$
$$\quad |\qquad |$$
$$(CF_2-CF)_n-O-CF_2-COOH$$

perfluoro sulfonic acid (nafion)

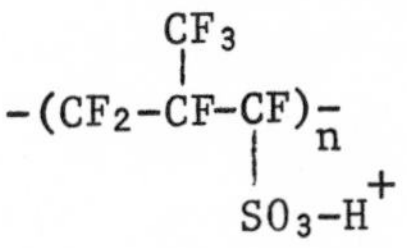

$$\qquad CF_3$$
$$\qquad |$$
$$-(CF_2-CF-CF)_n-$$
$$\qquad\ |$$
$$\qquad SO_3-H^+$$

radiation-grafted polyethylene (permion)

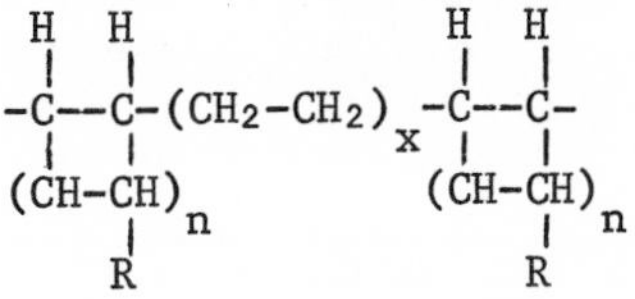

$$
\begin{array}{cc}
H\ \ H & H\ \ H \\
|\ \ \ | & |\ \ \ | \\
-C--C-(CH_2-CH_2)_x-C--C- \\
|\ \ \ | & |\ \ \ | \\
(CH-CH)_n & (CH-CH)_n \\
| & | \\
R & R
\end{array}
$$

condensation polymer

$$
\begin{array}{c}
O \\
\|\ \\
-O-(P-O-CH_2-CH_2)_n- \\
| \\
ONa
\end{array}
$$

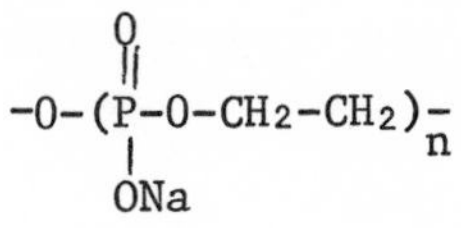

from lithium to titanium.

Cells with Se complexes, such as $Mg/Ba_xMg_{6-x}Se_6/NbSe$ are examples of how the electrolyte can adjust to accommodate the anode-metal cations (114b). Similar complexes can in principle be synthesized for Al, Ca and Ti. But Na and K complexes, such as Na_2SbO_3, $NaTaWO_6$, $2K_2O \cdot 3Nb_2O_6$, and some others with $K = n \times 10^{-5}$ ohm^{-1} cm^{-1} (114c), may also serve as solid electrolytes. Perhaps some with higher conductances will be found. Recent advances in synthetic inorganic chemistry raise the hopes of achieving some success in this direction. United efforts of electrochemists, crystallographers and inorganic chemists in the field of solid ionic conductors will increase the chances of creating HED SBs based on solid electrolytes.

Ion-exchanger resins as solid polymer electrolytes, impregnated with the cations of the chosen anode metal, may prove applicable. Their use in the fuel-cell/electrolyzer single module concept is already under investigation as to complexity and operability (115). Doubtless better SPE's will be discovered.

Moist and Mixed Electrolytes

We consider now the applicability of some water as a component in mixed aqueous-nonaqueous, or aqueous-molten salt electrolytes, and we deal with the possible use of the six light, active metals in SBs based on "moist" or "transition region" electrolytes.

The main reason for avoiding water as a solvent is the fact that the electrolysis of aqueous solutions of alkali and alkaline-earth metal salts commences at 1.7-2.0 volts (depending on the electrode material) and results in the evolution of O_2 and H_2. If the cell itself has a higher voltage, internal electrolysis can, but not always does occur, accompanied by the evolution of H_2 and O_2 and by self-discharge (117). However, this fact does not preclude attempts to create moist primary batteries with Li, Na or Ca, if the activity of H_2O is kept sufficiently low.

Minor admixtures of water to nonaqueous electrolytes are often harmful, for example in batteries with inorganic solvents such as $POCl_3$, $SOCl_2$, SO_2Cl_2, where it is important that the electrolyte be free of water contamination because of the possible formation of oxychloride cements:

$$MOCl_n + H_2O = MO(OH)Cl_{n-1} + HCl$$

where M = P or S (118). Special water scavengers (synthetic zeolites) have been proposed (119) to maintain the H_2O level in the electrolyte below that which interferes with the optimal performance of cells such as Li or $Na/POCl_3/(CF)_n$.

Primary Cell Status as Baseline. Some time ago a short
report was made on the Symposium on Batteries for Traction (see
(90)) on the studies in the Lockheed Research Laboratory on
electric power sources based on reactions of alkali metals with
water. These reactions produce very high ED's, are surprisingly
efficient: about 3000 Wh per kg of active metal (120).

Following this disclosure, several papers and patents
appeared which we have found to be very instructive (121-124):
under operating conditions the sometimes violently energetic
reactions can be controlled, especially when the activity of
water is kept low. However, conditions under which it is
possible to redeposit electrochemically the alkali metal in a
recharge require further study.

The claim of the authors of (123) that the electrochemical
behaviour of Li in aqueous solutions of its hydroxide is unique,
however, needs further examination. It has not yet been proved
experimentally that Na or Ca do not form films in concentrated
solutions, of the hydroxyl ion (OH^-). NaOH solutions can be
prepared in the so-called transition region between molten NaOH
and its aqueous solutions, i.e. with compositions at which the
mole function of water $\ll 0.2$. The physico-chemical properties
of the compositions in the transition region are quite different
from those of dilute solutions of salts or hydroxides in water.

Transition Region Considerations. The conductance of a
binary system can be approached from the values of conductivity
of the pure electrolyte; one follows the variation of conductance
as one adds water or other second component to the pure electrolyte. The same approach is useful for other electrochemical
properties as well: the e.m.f. and the anodic behaviour of light,
active metals, for instance. The structure of water in this
"transition region" (TR), and therefore its reactions, can be
expected to be quite different from its structure and reactions,
in dilute aqueous solutions. (The same is true in relation to
other non-conducting solvents.) The molecular structure of any
liquid can be assumed to be close to that of the crystals from
which it is derived. The narrower is the temperature gap between
the liquid and the solidus curve, the closer are the structures
of liquid and solid. In the composition regions between the pure
water and a eutectic point the structure of the liquid is
basically like that of water; between eutectic and the pure salt
or its hydrates the structure is basically that of these
compounds. At the eutectic point, the conductance-isotherm runs
through a maximum and the viscosity-isotherm breaks. Examples
are shown in (125).

Hydrates could play an important role in electrolytes for
batteries with active metals. The water, being involved in the
hydrate structure, is less active than at the compositions on the
water side of the diagram, i.e. between the eutectic and pure
water. The rate of the anodic dissolution of the alkali and

284   CHEMISTRY FOR ENERGY

alkaline-earth metals can be made low enough so that the anodic
reaction can be harnessed. Hydrates of the halogens, like Cl_2
can be employed for storage of the Cl_2 in a rechargeable battery.
   The ions of the salt, which are frozen in the lattice of the
solid salt, can move when the salt melts, and the melt will
conduct. The addition of a few percent of water lowers
substantially the viscosity (η) of the salt, acting as a
"lubricant" for the movement of the ions, increasing their
mobility and the conductivity of the melt. Thus, for example,
whereas for 100% $LiClO_3$ at 128°, K = 0.107 ohm^{-1} cm^{-1} and η = 66.1
cp, the addition of only 1.1% by weight (5.28 mole %) of H_2O
the conductivity is about 2 times that of pure $LiClO_3$ and η is
about 10 times less (126). The addition of small amounts of
water may lower substantially the melting point. Thus 14.5% of
water mentioned above decreases the m.p. of $LiClO_3$ from 127.6° to
20°C. For the binary system $LiNO_3$ (m.p. 260°) and NH_4NO_3 (m.p.
169°), the eutectic is at 97°. Other low-melting systems (see
Tables VI-VIII) can be formed at ambient temperatures (-10°C to
+50°C) by the addition of 5-10 wt% of H_2O to binary mixtures, but
the composition of the system at these temperatures will still
remain in the TR. If this composition can be controlled, electrodeposition may be possible for the light metals such as Na and
Ca. In the case of molten-salt electrolytes the addition of
water would lower the operating temperature towards ambient,
decreasing simultaneously the viscosity of the electrolyte and
increasing its conductivity.
   In the case of non-aqueous (inorganic and organic) solvents
the addition of small amounts of H_2O will sometimes increase the
usually poor conductivity, and permit the use of a comparatively
high-melting solvent. From such a binary electrolyte the
morphology of the metal deposited during recharge might be quite
acceptable. In Table XII we list some untried transition-region
electrolytes, all speculative, which might be considered.

TABLE XII. Some Untried Transition-Region Mixtures
            As Candidate SB Electrolytes

• water in molten KOH or molten oxy-salts

• moist salt in dioxane or acetamide

• dry HCl in urea

• trichlorosulfonate in anhydrous H_2SO_4

• moist salt in liquid NH_3

(-) active metal // electrolyte // oxidant (+)
       ?   ?   ?

Active Metals: Li, Na and Ca. The processes in the primary
Li batteries which were discussed in (120-124) cannot be reversed
electrochemically, because they are accompanied by the evolution
of H_2, which is removed from the cell. Experiments in which Li,
Na and K (127) are dissolved in aqueous solutions of their
hydroxides have shown a substantial decrease in the rate of the
process as the hydroxyl ion concentration is increased. Breaks
in the curve: rate-of-dissolution vs OH^--concentration, occur at
compositions corresponding to the formation of hydrates;
corresponding breaks were also observed on the conductivity- and
viscosity-composition curves. The decrease of the dissolution
rate is substantial. Thus when the concentration of NaOH was
increased from 7-m to 21-m, the rate decreased 200 times, and in
a solution containing 1-m H_2O in 10-m dioxane, the speed of Na
dissolution was reported to be a thousand times less than in pure
water (127). As to the $Cl_2 \cdot 6H_2O$ hydrate, it might be isolated
from the electrolyte by a separator, as is done in a semi-
aqueous Li-HgO primary battery (128), or replaced by another
cathode material, like Br_2, or $CuCl_2$ or other. A Li-alloy might
aid in achieving reversibility.

Let us consider some examples. $Li/LiNO_3 \cdot urea/C$, a primary Li
battery is patented in the U.S.S.R. (129a). A secondary battery
with an alkali metal or its alloy as anode, a strong oxidant
(strong acids, molten nitrates or nitrites, chlorates or
perchlorates) in an inert medium as electrolyte, and a cathode of
steel, Mo or other alloy, is described in a German patent granted
to the workers of the Moscow Energy Institute. It is claimed
that the violent reaction of the alkali metals with the oxidants
subsides if the heat produced in the reaction is conducted away.
Thus H_2SO_4, $H_2S_2O_7$, HSO_3Cl and HSO_3F are among the possible
oxidants listed (129b).

The battery system, $Li/H_2SO_4/Fe$ or Mo, has TED = 7.4 kWh/kg,
and the products of the reaction,

$$H_2SO_4 + 12Li = 2LiH + Li_2S + 4Li_2O$$

are mostly soluble in the liquid. The temperature of the anode
is 40-50° and the cathode (which is heated) is at 250-400°
(129b). Could it be made reversible with judicious addition to
the electrolyte?

The formation of an anodic film on alkali metal anodes is
mentioned in (129b). If it can be confirmed that Li is not
unique in its reaction with water, as it is claimed in (123-124),
then Na may also develop such a film in contact with H_2O or
non-aqueous liquids and so be protected. Design variations in
Na-H_2O primary batteries are described (130-132).

In all publications on alkali metal-water batteries in which
the cell reaction is disclosed, it is noted that hydrogen is
evolved. No SB can be based on this reaction as long as H_2 is
being lost. Perhaps it could be retained in the electrolyte

under high pressure. A more fruitful approach might be to react the anode with an oxidant like Cl_2, Br_2 or oxygen in such a way that the product formed remains inside the cell and H_2O does not decompose but works only as a lubricant for decreasing the viscosity, increasing the conductance and lowering the operating temperature. Thus in a cell such as $Me/MeAlCl_4/Cl_2 \cdot 6H_2O$, the metal chloride is the discharge product and Me and Cl_2 the charged reactants. Cl_2 is captured by H_2O to form $Cl_2 \cdot 6H_2O$. The $Zn/Cl_2 \cdot H_2O$ storage system is being developed (84).

If H_2 is evolved, it could be captured by the simultaneously deposited metal to form a hydride, although such a process would require special conditions. Na reacts with molten NaOH forming NaH at 400° (133). It would be worthwhile to study such reactions in cells at temperatures below 150°, using a NaOH + KOH eutectic to which a few percent of water has been added. Further, mixtures of KOH, NaOH, LiOH and of their eutectics, with water, acetone, dioxane (134) and other solvents, may serve as suitable electrolytes for SBs with sodium or other alkali metals. There are certainly interesting possibilities to be explored in this area.

Ca is cheap, and its sources are practically inexhaustible. Its reactions with oxidants have high theoretical ED's. The study of its behaviour in reversible cells might be rewarded by success. Several publications on the use of Ca in primary thermal batteries can be found in the literature. In one cell, with a Ca-Ba alloy and a molten salt electrolyte, the presence of water has been shown to be harmful (90a). However, detailed studies of the behaviour of a Ca anode in other mixed water and molten salt electrolytes have not been carried out. The reaction of Ba with water is slow, and alloying it in suitable proportion with Ca may be helpful in controlling the rate of the anodic dissolution of Ca. $Ca(OH)_2$ is very sparingly soluble in water, and film formation by it on the anodically polarized Ca surface is very probable. However, this film may dissolve rapidly enough in low-melting hydrates, such as $Ca(NO_3)_2 \cdot 4H_2O$ (m.p. 42.7°), $CaCl_2 \cdot 6H_2O$ (m.p. 30.2°), or $Mg(NO_3)_2 \cdot 6H_2O$ (m.p. 89.9°), to permit substantial anodic limiting-current densities, and to offer prospects of rechargeability. A cell of the type

$$Ca(Mg,Ba) \ \left| \ \begin{matrix} CaCl_2-AlCl_3 \\ OH^-, H_2O<5\% \end{matrix} \right| \ Cl_2 \cdot 6H_2O \text{ or } CaCl_2$$

might open the vistas for a useful SB based on an anode of Ca or one of its alloys.

Passive Metals: Mg, Al and Ti. Like Ca, Mg and Al are also cheap, and the TED of their couples with oxidants is high. A review of the performance of Mg and Al as primary cell anodes, published in 1959, indicated that Mg dry cells were already available for special purposes and might achieve commercial

significance, but that prospects for dry cells based on Al were
poor (135). Since 1960 several publications have been issued on
the behaviour of Mg, such as in aqueous solutions of $MgCl_2$ (136),
and on its use in aqueous primary batteries (1).

A comprehensive paper on the use of Mg alloys in aqueous power
sources has appeared recently. The author recommends more extensive use of Mg in primary batteries because the price per gram-
equivalent is 2.2 times smaller than that of Zn and the
theoretical ED is more than 8 times higher than that of Zn in acid
solution and about 6 times higher in basic solutions. Mg has the
most negative electrode potential that in practice can be achieved
in aqueous cells. The cell $Mg/NaClO_4/O_2$ is discussed (137a). The
metallurgy of the lithium-rich end of the Li-Mg dry cells commercially available are described in a recent monograph on primary
batteries (138). A commercial seawater-activated Mg battery is
described in (139) and there are several patents on that topic.

However, there are no known SB systems with Mg in aqueous
solutions. The Mg anode's irreversibility in aqueous solutions
is thought to be due, in part to the existence of monovalent Mg
ions during the electrochemical discharge, in part to the self-
corrosion and film formation, and in part caused by other factors
(136,140). All attempts to deposit this metal on the negative
electrode from aqueous electrolytes have failed. It is claimed
that the Mg cell with molten salt electrolyte, LiCl-KCl eut., is
reversible (141): it operates at temperatures above the eutectic
melting point, i.e. about 400°C. Small amounts of water might
decrease the operating temperature.

Mg cells with solutions of its salts in acetamide-H_2O
mixtures might show rechargeability. Thus, the interesting cell
$Mg/LiCl$-acetamide/V_2O_5 is mentioned in (1, p 100). Bromides of
Cu or Co might be more effective than V_2O_5. Once formed on the
surface, MgO is only slowly complexed away.

Since Al offers the possibility of a 2-to 3-fold reduction in
anode weight and volume over Zn in alkaline primary batteries,
many attempts have been made to use it. Cells of the type
Al/3M-KOH/C have been proposed for large Al-air or Al-O_2
batteries. Al alloys are also used, with porous Ni as cathodes
(142). There are several articles and patents dealing with
Leclanché-type dry cells in which Zn is replaced by Al and various
compounds are used as electrolytes. A cell $Al/HClO_4 + HCl/MnO_2$
or PbO_2, is offered in (143). Some organic compounds, like
HCO_2NH_2, are added to increase the shelf-life of the cell (144).
Water-activated reserve batteries with Al anodes are described,
the chlorine at the cathode being delivered by an organic compound, a commercially available chlorinating agent, which reacts
(145) according to the reaction:

$$C_3N_3O_3Cl_3 + 2Al + 6H_2O = C_3N_3O_3H_3 + 2Al(OH)_3 + 3HCl$$

Despite this work, in a recent book on primary batteries it is

stated that Al has not yet been utilized in a commercially
important battery and its realization still has to be considered
to be in the experimental stage (138). Wasteful anode corrosion
is still a problem.

None of these facts generates much hope for the realization
of rechargeable batteries with Al anodes in aqueous solutions.
However, the oxide film on Al can be complexed away in alkali
solutions. In very strong alkali, such as the low-melting
eutectics of KOH and NaOH, aluminum might be rechargeable. The
performance of Al in the TR of such a melt to which some water is
added also might be very interesting: the battery's operating
temperature could be markedly decreased and the film formed on
the Al rendered porous and perhaps reducible.

The possibilities for rechargeable Al systems based on other
moist inorganic and organic nonaqueous electrolytes are simply
not known.

The electrochemical properties of Ti were discussed in a
previous report and some pure and mixed electrolytes considered
(90b). The tight oxide film on Ti can be anodically dissolved in
acidic aqueous solutions (146,147). Electroplaters have found
complexes from which Ti can be deposited (148). However this
oxide film militates against both anodic oxidation and cathodic
reduction being accomplished in the same aqueous solution. No
passivation occurs during anodic dissolution of Ti in CH_3OH + HCl
in the absence of H_2O (149). Ti and Ti-alloy cathodes in aqueous
alkaline solutions absorb H and render the system rechargeable
(150-152). Thus cells based on $NiTi_2H_x$/strong $KOH/Ni(OH)_2$ have
been constructed and cycled (152).

The utilization of Ti as a rechargeable negative rests on the
attenuation of the effects of the passivating oxide film. In the
review, several dry molten salt systems are proposed which would
seem to have promise (90b). The effects of addition of moisture
to decrease both the m.p. and operating temperature were not
considered there.

Prospects for TR Electrolyte SBs. In view of the harmful
effects often cited in the literature of even small traces of
water on the operation of non-aqueous batteries with alkali metal
anodes, it might be supposed that electrolytes of the TR composition cannot be applied in such batteries. This same idea may
dominate when molten salt SBs are considered. Such a general
conclusion cannot be justified. A dilute solution of water in a
salt has the structure either of this salt proper or its adjacent
hydrate, and the energy, properties and reactions of this water
are quite different from those of pure water or of dilute
solutions of various compounds in it. On the other hand, a small
amount of water in the electrolyte system will decrease its
melting point and increase its conductivity. Mixtures of water
with such liquids as some alcohols or dioxane and other aprotic
and even proton-forming substances, may open new prospects for

non-aqueous SBs. Ref. (129b) may indicate a breakthrough.

Since the theoretical foundation of TR electrolytes is not applicable at this stage to the a priori selection of electrolytes for study, each new composition of such electrolytes must be examined experimentally with the prospective electrode materials of the battery. We should not continue to reject the transition region solutions as possible electrolytes for the active metals in SBs.

Perspective Drawn From the Review of Electrolytes for HED SBs

1. Conventional and Advanced Aqueous Electrolyte Systems
(a) Improvements are expected in all the factors that count:
     energy density (ED), cost/kWh, safety, reliability, as well
     as turnaround efficiency and cycle life at deeper depths of
     discharge (dod).
(b) Interesting chemistry is found in the investigation of new
     alloys, in the subtleties of dopants which decrease the
     polarizations, and in the development of new separator
     materials.
2. Dry Organic and Inorganic Electrolyte Systems
(a) Several new systems based on alkali metals and layered
     sulfides have very high energy density.
(b) In 1978 we still lack evidence of the arrival of the first
     rechargeable cell which is repetitively chargeable at greater
     than 60% depth of discharge at a practical rate of recharge.
(c) Interesting chemistry is found in the intercalation reactions
     of the layered compounds and in the barriers to rapid re-
     charge of the negative.
3. Molten Salt Electrolyte Systems
(a) The breakthrough was the Li-Al/LiCl-KCl eut./FeS system:
     intermediate ED, good cycle life proven at >60% dod. There
     are probably many other systems just as good or better.
(b) Interesting chemistry includes appearance of new crystalline
     complexes as intermediates; bewildering stability problems
     with separators, cases, seals; and very fast charge-transfer
     processes.
4. Solid Electrolyte Systems
(a) Good ionic conductors are emerging, both inorganic and
     organic, but the stability of open-lattice conductive paths
     as cycling proceeds seems doubtful, and solid state
     interface-integrity during cycling seems unlikely.
(b) Liq./solid/liq. systems offer best prospects, if solid
     electrolytes can be made thin, tough, and chemically re-
     sistant as well as conductive. Turnaround efficiency can be
     high, in both the $Na/\beta-Al_2O_3/S$ and the H_2/solid-polymer-
     electrolyte/Cl_2 systems, as class examples.
5. Transition Region (TR) ("Moist") Electrolyte Systems
(a) Low activity of water (or other second solvent) may permit
     cycling of light metals and permit oxide cathodes to be used.

Low-water chemistry could disclose interesting surprises.
(b) No system in the TR has been investigated in detail yet, from
the storage-battery viewpoint.

Who Wants Better Electrical Energy Storage?

Various electric power companies worldwide have analyzed their
needs for electric energy storage, and have projected cost
benefits (See (153), for example). In Canada the National Energy
Board and the Department of Energy, Mines and Resources have been
doing cost-effectiveness analyses of the storage of electrical
energy in times of plenty to be used in times of excess demand in
different geopolitical regions of Canada. In our Laboratory we
have focussed on a hypothetical new military base in the far
North; and for present purposes we shall use this work as an
example. The hypothetical base would be isolated from the grid
system, and fuel costs will be high (unless of course it is
located on top of a gas well site!). Relevant operating data on
the daily and seasonal variation in electricity demand for
isolated communities, are kept by the Northern Canada Power
Commission: from the quiet Eskimo village of Arctic Red River,
with 150 inhabitants, to the thriving northern center of Inuvik,
with 9000 inhabitants -- in powerplant terminology, of average
demand from 50 kW to 3.2 MW.
Results of this analysis (154) can be summarized as follows.
At a site the size of Cambridge Bay, population 700, where the
average seasonal demand is 510 kW and the diurnal variation is
very large (Figure 3), if storage capability were to be added to
power system, the diesel installation would need to have only 510
kW installed capacity instead of the 821 kW presently installed
to supply peak power loads. It would run at design load, where
the efficiency of use of the fuel is highest, and not at part
load, where the efficiency is considerably less. Credits would
accrue to lower cost of a smaller diesel installation and to
higher efficiency of its use of fuel. Debits would include the
costs of the storage-battery system and of the power-conditioning
system.
This operation was modelled using a sinusoidal representation
of the daily and seasonal power fluctuations. The Cost-Benefit
equation contains four terms:

$ Cost Benefit = Fuel Saving + Reduced Diesel Cost Benefit
- Cost of Power Conditioning - Cost of Storage Battery

each of which contains appropriate constant and variable parameters. In this work (154) the constants were determined from the
operational data. The total costs of operating with and without
storage were calculated and compared.
It was found that cost benefits could indeed be achieved if
storage were introduced, at today's Northern Canadian prices for

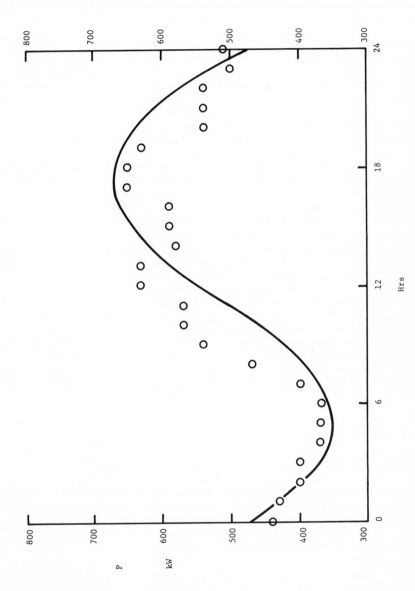

Figure 3. Hourly power demand Cambridge Bay: 1200 hr (6 Nov 75) to 1200 hr (7 Nov 75); (A) 510 kW, (D_2) 0.32

oil, batteries, power conditioning equipment: for a community in
which daily peak-trough cycling is high (no smoothing base loads),
i.e. when peak power is >20% above average power. Cost benefits
do increase with increasing difference between peak and average
power. Cost benefits are markedly dependent upon the life style
of the community, it seems.

The numbers generated put into quantitative terms the cost
benefits which can accrue from storage if it is applied properly.
They give a much firmer basis upon which to plan an experimental
demonstration of benefits of electrical energy storage to a small
community. This could be a harbinger of things to come in larger
communities if fuel costs and electricity demand continue to
soar, and if peak-power loads are allowed to increase.

The general conclusion is that with battery storage at
selected locations one could save high-priced hydrocarbon fuel.
This is not to say that one could, through storage, circumvent
the second law of thermodynamics: every energy transformation
exacts its penalty, increases the entropy of the system
somewhere. The choice depends upon WHAT one wants to save, and
WHERE one wants to save it. A saving transformation in one
location will assuredly cause a dissipation somewhere else in the
total energy system.

The prospects for greater investment in electrical-energy
storage are being assessed world-wide. In many places, Canada
included, it is no longer a trivial concern. Conservation is the
goal which perfuses these analyses. A reminder which we have
found useful in such assessments, and which the reader may find
helpful also, is offered in Figure 4. The symbols are standard.
The last equation is provocative: economic losses can be equated
to entropy losses only if the total energy system under consideration is the whole universe; for simpler, practical "total energy"
systems, cost-accounting is definitive, the equivalance usually
does not hold, and the assumption that it does can be highly
misleading.

Industrial Viewpoints on R & D in Electrical Energy Storage

The Canadian Battery Manufacturers Association recently
completed a study called "Electrical Energy Storage Research
Plan" (155). They made considerable effort to underscore the
impediments to industrial research being done in Canada under
federal government support, citing particularly the loss of
industrial security necessary in a free-enterprise society, the
need to assign the patent rights to the Crown, and the lack of
continuity or 5-year planning in the proposed federal program.
Some multinationals in this field find it less costly and less of
a legal hassle to do their R & D elsewhere than in Canada.
Others consider the Canadian programs for assistance to
technological innovation to be a model which other governments
might well emulate. The new tax incentives announced April 78

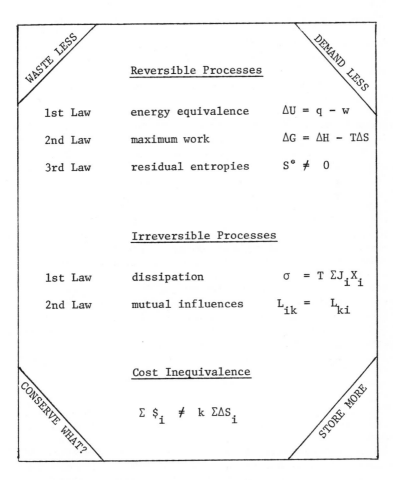

Figure 4. Conservation axioms

294   CHEMISTRY FOR ENERGY

make the situation even more attractive, and will promote more
industrial R & D in this technical area.

The second significant thing the CBMA did was to put into an
order of priority the technical topics on which they believe
federal R & D money, and their own, ought to be spent. Electric
vehicle batteries headed the list of the manufacturers' priority
topics. Apparently they, like others, can foresee the
introduction of electric vehicles on a massive scale in Canada
during the period 1980-1995.

Members of the Canadian Institute of Mining and Metallurgy,
and of the Canadian Manufacturing Association have become very
concerned about the costs of energy and the potential of load-
levelling in industrial processing. The electrometallurgy and
chlor-alkali industries in Canada contribute a much higher
fraction to the Gross National Product of Canada than do these
two industries in most other countries, and the storage and more
flexible use of electricity industrially is becoming more important even in a country where electricity costs have traditionally
been low. Electrochemical storage systems -- storage batteries
of yet new and different dimensions -- have become a much more
fashionable topic in the considerations of plant renovation or
new-plant design in Canada.

Finally, the crown corporation Atomic Energy of Canada Ltd.,
which is responsible for the development of nuclear energy
sources in Canada, and the hydroelectric power companies, which
are provincially (publicly) owned utilities, face the challenge
of determining what kind and how much energy-storage capability
to build into their source/grid systems. These needs and
opportunities vary widely from one geographic area to another.
The federal Department of Energy, Mines and Resources assists in
providing the technical aid they need, and the Department of
Industry, Trade and Commerce offers financial and analytical
stimuli. Indeed, we conclude that, selected increased effort on
research, development and demonstration in this field seems to be
well justified.

Terms. TED - theoretical energy density (free energy of
reaction/sum of molar wts of reactants); ED practical or realized
Wh/kg; SB - secondary or storage battery (rechargeable);
dod - depth of discharge (% recharge removed before recharge);
TR - transition region: nearly pure (97-100%) ionic conductor
mixed with second compound; K - specific conductivity ohm^{-1} cm^{-1};
D.C. - dielectric constant; η - viscosity, centipoises;
R & D - research and development ΔG, ΔH, ΔS, q, w - classical
thermodynamic significance; J, X, L - fluxes, forces and
phenomenological coefficients of irreversible thermodynamics;
D - diffusion coeff., cm^2/sec.; SPE - solid polymer electrolyte;
PEO - polyethylene oxide; AN, THF and other solvents: Table II.

Literature Cited

1. Jasinski, R. "High Energy Batteries", Plenum Press, N.Y. (1967).
2. Jasinski, R., Electrochem. Technol., 6, 28 (1968).
3. Klochko, M.A., J. Appl. Chem. (Russ.) 9, 420 (1936).
4. Jorne, J. and Tobias, C.W., J. Appl. Electrochem., 5(4), 279 (1975).
5. Auborn, J.J. et al., Proc. 25th Power Sources Symp., p. 6 (1972).
6. Auborn, J.J. et al.
    a. Electrochem. Society Extended Abstr., Vol. 73-2, Abstr. No. 58, p. 143 (1973).
    b. J. Electrochem. Soc, 120, 1613 (1973).
7. Auborn, J.J. et al., Proc. 26th Power Sources Symp., p. 45 (1974).
8. Auborn, J.J. and Marincic, N., "Power Sources 5", p. 683 (1975).
9. Auborn, J.J., U.S. Patents 3,891,157 and 3,897,264; Chem. Abstr., 83, 134971 and 150351 (1975).
10. Auborn, J.J. et al., Chem. Abstr., 84, 138293 (1976).
11. Marincic, N.
    a. Proc. 26th Power Sources Symp., p. 51 (1974).
    b. Extended Abstr., Vol. 75-1, Abstr. No. 25, p. 56 (1975).
    c. Marincic, N. and Epstein, J., J. Electrochem. Soc., 123, 68c, Abstr. No. 9 (1976).
12. Marincic, N., U.S. Patent 3,907,593, Sept. 23, 1975; Chem. Abstr., 84, 20259 (1976).
13. Behl, W.K. et al., J. Electrochem. Soc., 120, 1619 (1973); Extended Abstr., Vol. 73-2, Abstr. No. 59, p. 145 (1973).
14. Driscoll, J.R. and Holleck, G.L., Extended Abstr., Vol. 75-2, Abstr. No. 34, p. 91 (1975).
15. Dey, A.N., J. Electrochem. Soc., 123, 68c, Abstr. No. 8 (1976).
16. Miles, M.H. and Harris, W.S., J. Electrochem. Soc., 121, 459 (1974).
17. Elzinga, C.H. and Vermeulen, C.G., "Power Sources 5", 603 (1975).
18. Keller, R. and Evans, S., J. Electrochem. Soc., 114, 655 (1967).
19. Besenhard, J.O. and Eichinger, G., "High Energy Density Lithium Cells. Part I: Electrolytes and Anodes", J. Electroanal. Chem., 68, 1-18 (1976).
20. Walden, Paul, "Elektrochemie nichtwässeriger Lösungen", Leipzig, 515 (1924).
21. Kessler, Yu.M. et al., "Electrochemical Properties of Aluminum Compounds in Nonaqueous Solutions", Advances in Chemistry, (Russ.), 33, 261 (1964).
22. Gilman, S., Proc. 26th Power Sources Symp., 28 (1974).
23. Gunther, R.G., "Power Sources 5", 729 (1975).
24. Dey, A.N., Electrochim. Acta, 21, 377 (1976).

25. Tiedemann, W.H. and Bennion, D.N., J. Electrochem. Soc., 120, 1624 (1973).
26. Wilburn, N.T., Proc. 25th Power Sources Symp., 3 (1973).
27. Gabano, J.P. et al., J. Electrochem. Soc., 119, 459 (1972).
28. Levy, S.C., Extended Abstr., Vol. 72-2, Abstr. No. 1, p. 9 (1972).
29. Brooks, E.S., Proc. 26th Power Sources Symp., 31 (1974).
30. Lang, M. et al., Proc. 26th Power Sources Symp., 37 (1974).
31. Lehmann, G. et al., "Power Sources 5", 695 (1975).
32. a. Bro, P. et al., "Power Sources 5", 703 (1975).
     b. Nagy, G.D., "Lithium Batteries - An Overview", DREO Technical Note No. 74-29 (1974).
33. Fukuda, M. and Iijima, T., "Power Sources 5", 713 (1975).
34. a. Morita, A. and Iijima, T., Jap. Patent 75, 145,817, November 22, 1975; Chem. Abstr., 84, 138329 (1976).
     b. Eda, N. and Iijima, T., Jap. Patent 75, 136,631, October 30, 1975; Chem. Abstr., 84, 153111 (1976).
35. Dey, A.N., J. Electrochem. Soc., 118, 1547 (1971).
36. Nicholson, M.M., J. Electrochem. Soc., 121, 737 (1974).
37. Kato, H. and Saito, T., Jap. Patent 74, 105,927, October 7, 1974; Chem. Abstr., 83, 13320 (1975).
38. Toyoguchi, Y. and Iijima, T., Jap. Patent 75, 145,816, November 22, 1975; Chem. Abstr., 84, 138330 (1976).
39. Toyoguchi, Y. and Iijima, T.
     a. Jap. Patent 75, 52,530, May 10, 1975; Chem. Abstr., 83, 182065 (1975).
     b. Jap. Patent 75, 52,525, May 10, 1975; Chem. Abstr., 83, 182066.
40. Zotov, N.L. and Khlystova, N.B., Elektrokhimia, 2, 1444 (1966).
41. Makarenko, B.K. et al.
     a. Elektrokhimia, 9, 705 (1973).
     b. Elektrokhimia, 10, 355 (1974).
42. Caiola, A. et al., Electrochim. Acta, 17, 421 (1972).
43. a. Holleck, G.L. and Driscoll, J.R., Extended Abstr., Vol. 75-2, Abstr. No. 32, p. 86 (1975).
     b. Whittingham, M.S., Science, 192, No. 4244, 1126 (1976).
44. Murphy, D.W. and Trumbore, F.A.
     a. Extended Abstr., Vol. 75-1, Abstr. No. 26, p. 57 (1975).
     b. J. Electrochem. Soc., 123, 960 (1976).
45. a. Broadhead, J. and Trumbore, F., "Power Sources 5", 661 (1975).
     b. Bro, J., "Power Sources 5", 680 (1975).
     c. Marincic, N., "Power Sources 5", 681 (1975).
46. Broadhead, J. et al., U.S. Patent 3,864,167, February 4, 1975; Chem. Abstr., 83, 63393 (1975).
47. Chianelli, R.R. et al., Extended Abstr., Vol. 75-2, Abstr. No. 31, p. 84 (1975).
48. Whittingham, M.S., J. Electrochem. Soc., 123, 315 (1976); 124, 1387 (1977).
49. Walk, C.R. and Gore, J.S., Extended Abstr., Vol. 75-1, Abstr. No. 27, p. 60 (1975).

50. a. Rauh, R.D. et al., Extended Abstr., Vol. 76-1, Abstr.
        No. 6, p. 19 (1976).
    b. Brummer, S.B. et al., Electrochem. Soc. Proc., Vol. 77-6,
        p. 975 (1977).
51. Weininger, J.L. and Secor, F.W., J. Electrochem. Soc., 121,
    315 (1974).
52. Gore, J.S. and Walk, C.R., U.S. Patent 3,929,504, December
    30, 1975; Chem. Abstr., 85, 23559 (1976).
53. Broadhead, J. et al., U.S. Patent 3,928,067, December 23,
    1975; Chem. Abstr., 85, 80843 (1976).
54. Winn, D.A. and Steell, B.C.H., Extended Abstr., Vol. 75-1,
    Abstr. No. 22, p. 51 (1975).
55. Will, F.G. and Hess, H.J., German Patent 2,347,600,
    April 11, 1974; Chem. Abstr., 81, 52241 (1974).
56. Kawai, T., Jap. Patent 76, 38,030; 38,031; 40,527; 40,528;
    March, 30, 1976 - April 5, 1976; Chem. Abstr., 85, 80865-
    80868 (1976).
57. Caiola, A. et al., Electrochim. Acta, 17, 1401 (1972).
58. "The XIth Mendeleev Congress - A Review of the Latest
    Achievements of the (Soviet) Chemical Science and Technology",
    an Editorial, J. Appl. Chem. (Russ.), 49(2), I (1976).
59. Mann, C.K., Electroanal. Chem., 1, 58 (1969).
60. Klochko, M.A. and Mikhailova, M.P., J. Inorg. Chem. (Russ.),
    5, 2319 (1960).
61. Klochko, M.A. and Gubskaya, G.F., Izv. Sect. Fiz. Khim.
    Analiza, 27, 393 (1956); J. Inorg. Chem. (Russ.), 3, 2375,
    2571 (1958).
62. Kuznetsov, S.I. and Vaulin, L.V., Izvestia VUZ, Tsvetnaia
    Metallurgia, 5, 49 (1972); 5, 41 (1973).
63. Lorenz, R., Elektrochemie Geschmolzenen Salze, Vol. 1-3 (1906-
    1909).
64. Delimarsky, Yu.K. and Markov, B.F., "Electrochemistry of
    Molten Salts", Gosudarst. Nauch.-Tech. Izdatel. Lit..., 1960.
65. Antipin, L.N. and Vazhenin, S.F., "Electrochemistry of
    Molten Salts", Metallurgizdat., Moscow, 1963 (Russian).
66. Blander, M., "Molten Salt Chemistry", Interscience (Wiley),
    N.Y. (1964).
67. Brumeaux, M. et al., Electrochim. Acta, 13, 1591 (1968).
68. Swinkels, D.A.J., Chapter 4 in "Advances in Molten Salt
    Chemistry", A. Braunstein, Gleb Mamantov and G.P. Smith,
    Editors, Plenum Press, N.Y., 1971, pp 165-223.
69. Cairns, E.J. et al., "Cells - High Temperature", in
    Kirk-Othmer's "Encyclopedia of Chemical Technology", Second
    Edition, Supplement Volume, p. 120-137 (1971).
70. Cairns, E. and Steunenberg, R., "High Temperature Batteries",
    in "Progress in High Temperature Physics and Chemistry",
    Vol. 5, C.A. Rouse - Editor, p. 63-124 (1973).
    a. Ukshe, E.A. and Bukun, N.G., "The Dissolution of Metals
        in Molten Halogenides", Advances in Chemistry (Russ.),
        30, 243 (1961).

71. Bradley, T.G. and Sharma, R.A., Proc. 26th Power Sources Symp., 60 (1974).

72. Shimotake, H. et al., Extended Abstr., Vol. 73-2, Abstr. No. 13, p. 38 (1973).

73. Gay, E.C. et al., Chem. Abstr., 85, 35453 (1976).

74. Sharma, R.A., J. Electrochem. Soc., 123, 448 (1976).

75. a. Nelson, P.A. et al., Proc. 26th Power Sources Symp., 65 (1974).

    b. Shimotake, H. and Walsh, W., Chem. Abstr., 85, 65621 (1976).

    c. Steunenberg, R.K. and Roche, M.F., Electrochem. Soc. Proc., Vol. 77-6, p. 869 (1977).

76. McCoy, L.R. et al., Proc. 26th Power Sources Symp., 68 (1974).

77. L'vov, A.L. et al., Elektrokhimia., 11, 1322 (1975).

78. Yao, N.R. et al., Extended Abstr., Vol. 74-2, Abstr. No. 58, p. 140 (1974).

79. Caulder, S.M. and Simon, A.C., Extended Abstr., Vol. 74-2, Abstr. No. 56, p. 136 (1974).

80. Selover, T.B. Jr., Extended Abstr., Vol. 75-1, Abstr. No. 30, p. 65 (1975).

81. Frost, W.R. et al., Extended Abstr., Vol. 76-1, Abstr. No. 20, p. 51 (1976).

82. James, S.D. and DeVries, L.E., J. Electrochem. Soc., 123, 321 (1976).

83. a. Clark, R.P. and Grothaus, K.R., J. Electrochem. Soc., 118, 1680 (1971).

    b. Bush, D. and Baldwin, A., "Power Sources 5", No. 35, 581 (1975).

84. Preto, S.J., Ross, L.E., Martin, A.E. and Roche, M.F., Proc. Symp. and Workshop on Advanced Battery Research and Design, ANL76-8, Argonne National Lab., Illinois, 22-24, 1976.

85. a. Demidov, A.I. and Morachevskii, A.G., U.S.S.R. Patent 485,515, September 25, 1975; Chem. Abstr., 84, 33576 (1976).

    b. Werth, J.J., German Patent 2,348,258, March 28, 1974; Chem. Abstr., 81, 66192 (1974).

86. Buzzelli, E.S., U.S. Patent 3,650,834, March 21, 1972; Chem. Abstr., 76, 148255 (1972).

87. Mamantov, G. et al., Extended Abstr., Vol. 74-2, Abstr. No. 8, p. 24 (1974); Proc. Symposium and Workshop on Advanced Battery Research and Design, ANL76-8, p. B-152 (1976).

88. Redey, L. et al., "Power Sources 5", 559 (1975).

89. Anufrieva, N.I., J. Appl. Chem. (Russ.), 44, 1768 (1971).

90. a. Gross, S., "Candidate Batteries for Electric Vehicles", in Proc. Symp. on Batteries for Traction and Propulsion, p. 9-25, March 1972.

    b. Klochko, M.A. and Casey, E.J., J. Power Sources, 2, 201-232 (1977/78).

91. Casey, E.J. and Dubois, A.R., Can. J. Chem., 49, 2733 (1971).

92. Boston, C.R., "Molten Salt Chemistry of the Haloaluminates", Chapter 3 in "Advances in Molten Salt Chemistry" (See Ref. 68), p. 129-164.

93. Janz, G.J., "Molten Salt Handbook", Acad. Press, New York (1967).

94. Franzosini, P. and Ferioni, P., "Molten Salts with Organic Anions", An Atlas of Phase Diagrams, Pavia, Italy (1973).

95. Foley, R.T., J. Electrochem. Soc., 116, 13c-21c (1969).

96. Ukshe, E.A. and Buokun, N.G., "The Problem of Solid Electrolytes", Elektrokhimia, 8, 163 (1972).

97. a. Owens, B.B. and Argue, G.R., Science, 157(3786), 308-10 (1967); Adv. Electrochem. Eng., 8, 1-62 (1971).
    b. Takahashi, T. et al., J. Electrochem. Soc., 116, 357 (1969); 117, 1 (1970); 120, 1607 (1973). J. Appl. Electrochem., 2, 51 (1972); 3, 23 (1973).
    c. Brodd, R.J. et al., J. Electrochem. Soc., 125, 271c (1978).

98. Owens, B.B. et al., J. Electrochem. Soc., 118, 1145 (1971).

99. a. Ikeda, H. et al., Jap. Patent 75, 107,431, August 23, 1975; Chem. Abstr., 84, 138307 (1976).
    b. Margalit, N., Extended Abstr., Vol. 76-1, Abstr. No. 7, p. 21 (1976).

100. a. Shirokov, Yu.V. et al., Elektrokhimia, 8, 579 (1972).
     b. Bondarchuk, L.I. et al., Elektrokhimia, 11, 1325 (1975).

101. a. Ivanov, V.E. et al., Elektrokhimia, 8, 842 (1972).
     b. Ivanov, V.E. et al., Elektrokhimia, 11, 1418 (1975).

102. Ivanov, V.E. et al., J. Appl. Chem. (Russ.) 47, 670 (1974).

103. Weber, N. and Kummer, J., Proc. 21th Power Sources Symp., 37 (1967).

104. Yao, Y.Y. and Kummer, J.T., J. Inorg. Nucl. Chem., 29, 2453 (1967).

105. Kennedy, J.H. and Sammels, A.F., J. Electrochem. Soc., 121, 1 (1974).

106. Sudworth, J.L. and Tilley, A.F., German Patent 2,401,726, July 25, 1974; Chem. Abstr., 83, 63378 (1975).

107. a. Bones, R.J. et al., "Power Sources 5", No. 32, 539-557 (1975).
     b. Wiener, S.A. and Tischer, R.P., Annual Reports by Ford Motor Co. on Contract NSF-C805 (AER-73-07199), 1974-1977.
     c. Lazennec, Y. et al., J. Electrochem. Soc., 122, 734 (1975).

108. "New Scientist", p. 77, July 11, 1974.

109. a. Scrosati, B., J. Appl. Electrochem., 2, 231 (1972).
     b. Scrosati, B. and Torroni, M., Electrochim. Acta, 18, 225 (1973).
     c. Schneider, A.A. et al., "Power Sources 5", 651 (1975).

110. Heed, B. and Lundén, "Power Sources 5", 573 (1975).

111. a. Bukun, N.G. et al., Elektrokhimia, 10, 677 (1974).
     b. Pivnik, E.D. et al., U.S.S.R. Patent 190,212, October 30, 1975; Chem. Abstr., 84, 47152 (1976).

112. Danilov, A.V. et al., J. Appl. Chem. (Russ.), 47, 2053
     (1974).
113. Morachevskii, A.G. and Bikina, G.V., J. Appl. Chem. (Russ.),
     49, 258 (1976).
114. a. Raistrick, I.D. et al., Extended Abstr., Vol. 75-1,
        Abstr. No. 21, p. 50 (1975).
     b. Butherus, A.D. et al., Extended Abstr., Vol. 72-2,
        Abstr. No. 11, p. 35 (1972).
     c. Singer, J. et al., J. Electrochem. Soc., 123, 614 (1976).
115. a. Gileadi, E., Srinivasan, S., Salzano, F.J., Braun, C.,
        Beaufrère, A., Nuttal, L.J. and Laconti, A.B., J. Power
        Sources, 2, 191 (1977/78).
     b. Beaufrere, A., Yeo, R.S., Srinivasan, S., McElroy, J.M.
        and Hart, G., Proc. 12th IECE Conf., 959 (1977).
116. Bianchi, G. and Mussini, T., Ric. Sci. Rend., 6A, 37 (1964).
117. Bagotsky, V.S., "New Electrochemical Sources of Power"
     (A Review), Vestnik Akad. Nauk SSSR, (Herald of the U.S.S.R.
     Academy of Sci.) No. 7, p. 41-52 (1976).
118. French, K. et al., J. Electrochem. Soc., 121, 1045 (1974).
119. Casey, John E. and Chireau, R.F., U.S. Patent 3,864,168,
     February 4, 1975; Chem. Abstr., 83, 63392 (1975).
120. Roll, R.R., "Lockheed Electric Power Sources", in "Proc.
     Symp. Batteries for Traction" (See Ref. 90) 209 (1972).
121. a. Halberstadt, H.J. and Rowley, L.S., German Patent
        2,507,396, August 21, 1975; Chem. Abstr., 83, 166903
        (1975).
     b. Bauman, H.F., Chem. Abstr., 84, 167271 (1976).
     c. Halberstadt, H.J., Chem. Abstr., 85, 65620 (1976).
122. Littauer, E.L. and Tsai, K.C.
     a. Proc. 26th Power Sources Symp., 57 (1974).
     b. Extended Abstr., Vol. 74-2, Abstr. No. 59, p. 142 (1974).
     c. Littauer, E.L. and Redlien, J.J., Chem. Abstr., 84,
        167261 (1976).
123. Littauer, E.L. and Tsai, K.C.
     a. J. Electrochem. Soc., 123, 771 (1976).
     b. J. Electrochem. Soc., 123, 964 (1976).
124. Bennion, D.N. and Littauer, E.L., J. Electrochem. Soc., 123,
     1462 (1976).
125. Klochko, M.A., "On the Interrelation Between the Compositions
     of Conductance Maxima and of the Eutectic Point in Salt-
     Water Systems", Doklady, Adad. SSSR, 82, 261 (1952).
126. Klochko, M.A. and Grigoriev, I.G., "Conductivity and
     Viscosity of the System: $LiClO_3-H_2O$", Izv. Sect. Fiz-Khim.
     Analyiza, 21, 288 (1952).
127. Miskinova, T.A. et al., J. Phys. Chem. (Russ.), 46, 1000
     (1972).
128. Pistoia, G. and Marinelli, M., J. Appl. Electrochem., 2(2),
     157 (1972).
129. a. Antropov, L.I. et al., U.S.S.R. Patent 474,872,
        June 25, 1975; Chem. Abstr., 83, 166882 (1975).

129. b. Lobanov, A.A. et al., German Patent 2,316,979,
          November 22, 1973; Chem. Abstr., 80, 55349 (1974).
130. Mitoff, S.P. and Will, F.G., U.S. Patent 3,703,415,
      November 21, 1972; Chem. Abstr., 78, 51783 (1973).
131. a. Fustier, D.A., Graydon, W.F. and Foulkes, F.R., J.
          Electrochem. Soc., 123, 1259 (1976).
     b. Goldberg, B.D. and Foulkes, F.R., J. Electrochem. Soc.,
          124, 1819 (1977).
132. Will, F.G. and Mitoff, S.P., U.S. Patent 3,833,422,
      September 3, 1974; Chem. Abstr., 83, 63382 (1975).
133. Vorob'ev, T.A. et al., J. Inorg. Chem. (Russ.), 17, 2330
      (1972).
134. Klochko, M.A. and Godneva, M.M., J. Inorg. Chem. (Russ.),
      4, 2127, 2136, 2347, 2354 (1959).
135. Glicksman, R., J. Electrochem. Soc., 106, 457 (1959).
136. a. Casey, E.J. and Bergeron, R.E., Can. J. Chem., 40, 463
          (1962).
     b. Brossard, L., Thesis, U. de Montréal (1978).
137. a. Wisner, K. et al., "Power Sources 5", 425-446 (1975).
     b. Atkinson, J.T. and Sahoo, N., private communication.
138. Cahoon, N.C. and Heise, G.W. (Editors), "The Primary
      Battery", Vol. II, John Wiley and Sons, p. 528 (1976).
139. Faletti, D.W., J. Electrochem. Soc., 120, 1145 (1973).
140. Hoey, G.R. and Cohen, M., J. Electrochem. Soc., 105, 245
      (1958).
141. Selis, S.M. et al., J. Electrochem. Soc., 106, 135 (1959).
142. Zaromb, S., J. Electrochem. Soc., 109, 1125 (1962).
143. Kuzmina, A.V., J. Appl. Chem. (Russ.), 43, 898 (1970).
144. Lyakh, O.D. et al., U.S.S.R. Patent 338,949, May 15, 1972;
      Chem. Abstr., 77, 82977 (1972).
145. Gilson, A. et al., "Power Sources 5", 447 (1975).
146. a. Thomas, N.T. and Noble, K., J. Electrochem. Soc., 116,
          1748 (1969).
     b. Armstrong, E.D. et al., J. Electrochem. Soc., 117, 1003,
          (1970).
     c. Mandry, M.J. and Rosenblatt, G., J. Electrochem. Soc.,
          118, 29 (1971).
     d. Cowling, R.D. and Hintermann, H.E., J. Electrochem. Soc.,
          117, 1447 (1970).
147. Wenger, G. et al., German Patent 2,124,935, November 30,
      1972; Chem. Abstr., 78, 51780 (1973).
148. a. Kudriavtsev, N.T. et al., Doklady Akad. Nauk. SSSR, 132,
          636 (1960); 146, 1339 (1963).
     b. Pervii, E.N. et al., Ukrain. Khim. Zh., 39, 553 (1973).
149. Mansfield, F., J. Electrochem. Soc., 118, 1412 (1971).
150. Dobrov, Yu.V. et al., J. Appl. Chem. (Russ.), 47, 2115
      (1974).
151. a. Beccu, K. and Siegert, R., French Patent 1,568,808,
          May 2, 1969; Chem. Abstr., 72, 27790 (1970).
     b. Brit. Patent 1,276,790, June 1, 1972; Chem. Abstr., 77,
          82950 (1972).

152. Lidorenko, N.S. et al., Elektrokhimia, 10, 808 (1974).
153. Kalhammer, F., Am. Chem. Soc., Div. Fuel Chem., Prepr. 19(4),
     56 (1974); Proc. Symp. on Energy Storage, Electrochem. Soc.,
     p. 1 (1976).
154. Adams, W.A., Gardner, C.L. and Casey, E.J., "Electrochemical
     Energy Storage Systems: A Small Scale Application to
     Isolated Communities in the Canadian Arctic", DREO Memorandum
     No. 26/78 (ECD), May 1978.
155. "Electrochemical Energy Storage Research Plan for 1980-1995",
     Canadian Battery Manufacturers Association, on Contract No.
     2SR77-00172 to Dept. of Supply & Services, April 1978.

RECEIVED September 25, 1978.

Fuels Cells—Their Development and Potential

GUY BÉLANGER

Institut de Recherche de l'Hydro-Québec, Varennes, P. Q.

A fuel cell is a device that converts the free energy change of a chemical reaction directly into electrical energy. This conversion occurs by two electrochemical half cell reactions. This conversion is not subject to the Carnot cycle limitations and is thus theoretically more efficient than a heat-based process.

In this paper, we will discuss the thermodynamic principles involved in fuel cells as well as the kinetic aspects of their half cell reactions. In the kinetic considerations, we will also touch, briefly, on the fundamental problem of electrocatalysis. We will then proceed to describe different types of fuel cells and finally present the status of this new electrical generation device. Very recently, Kordesch (1), presented an excellent and detailed historical review of the past 25 years of fuel cell development. In that review (1), a description of practically all devices built and tested in the U.S. and Europe was presented. In this paper, the emphasis will be placed on the fundamental principles as well as the different factors that limit the fuel cell; we will discuss the most recent development of stationary power sources.

The basic principles were stated more than 100 years ago by Sir William Grove in 1839 (2) but the first real application was made in the American space flights of Gemini and the Apollo moon trips. After more than a decade of intense activity in fuel cell R and D, the commercial availability of these units is now in sight

I. Thermodynamic Considerations

As stated above, the fuel cell will convert the free energy change of a chemical reaction into electricity. In Table I we illustrate the different thermodynamic values for a typical fuel cell reaction, namely the hydrogen-oxygen reaction. The ΔH° of the reaction at 25°C is -68.32 kcal mole^{-1}; however the useful work that we can extract is the ΔG° (the free energy change) and this value is -56.69 kcal mole^{-1}; the difference between these values is the entropy change and represents a thermal loss. The maximum electrical efficiency, ε, that can be obtained from any chemical reaction is given by the ratio of the ΔG and ΔH of that reaction.

Table I: Thermodynamics Data for the hydrogen-oxygen fuel cell

$$H_2 \text{ gas} + 1/2 \ O_2 \text{ gas} \rightarrow H_2O \text{ liquid}$$

$$\Delta H° = \Delta G° + T\Delta S°$$

$$\varepsilon = \frac{\Delta G°}{\Delta H°}$$

$$\Delta H° = -68.32 \text{ kcal mole}^{-1} \equiv 1.481 \text{ V}$$

$$\Delta G° = -56.69 \text{ kcal mole}^{-1} \equiv 1.229 \text{ V}$$

$$\varepsilon = 0.830$$

In the above example ε is equal to 0.830. The ΔG of the overall
reaction can be expressed in terms of the corresponding battery
voltage and for the hydrogen-oxygen reaction at 25°C, its value is
1.229 V. As the temperature increases this thermodynamic equilibrium value will decrease by a factor of 0.84 mV per °C. If the
water produced remains in the gas phase, the ratio of $\Delta G°/\Delta H°$
increases to 0.911. So as we see, these values are much higher
than what can be obtained by a heat engine where the efficiency
is defined by the ratio of the temperature difference of the hot

and cold sources to the temperature of the hot source: $\varepsilon = \dfrac{T_h - T_c}{T_h}$

The efficiency of 1 can hypothetically be achieved at the
absolute zero for the heat sink; in practice T_c is around 273°K and
the T_h value can be increased to a practicable desired value.
      In Table II we list a series of reactions with either $\Delta H°$,
$\Delta G°$ and the maximum theoretical electrical energy efficiency
values. We may notice the high ε values for most reactions and for
some even a thermal efficiency of over 100%. This rather surprising result arises due to the positive value of the entropy change
of the reaction ($T\Delta S$ term) concerned. The maximum values pertain
to the situation of no load. As we may easily anticipate, the
real world is much different and a real device will not achieve
these terrific values under normal conditions of operation. These
limitations arise from kinetic factors and we will briefly outline
them next.

II. Kinetic Considerations

      a) Reversible potential. As stated above, at equilibrium, the
potential of a fuel cell should correspond to the $\Delta G°$ of the
reaction: for H_2/O_2 at 25°, two electrodes dipped in a conducting
electrolyte at 1 atm. H_2 and 1 atm. O_2 should read 1.229 V on a
high-impedance voltmeter. However, the potential observed is
closer to 1 V than to 1.229 V. There has been quite an extensive

TABLE II : Thermodynamic data for typical fuel cells electrochemical reactions

Reaction	$\Delta G°$	$\Delta H°$	$\epsilon = \Delta G°/\Delta H°$
25 °C	volts	volts	
$H_2 + 1/2\ O_2 \rightarrow H_2O$	1.229	1.481	.830
$CH_3OH + 3/2\ O_2 \rightarrow CO_2 = 2H_2O$	1.214	1.255	.967
$N_2H_4 + O_2 \rightarrow N_2 + 2H_2O$	1.559	1.612	.967
$CH_4 + 2O_2 \rightarrow CO_2 + 2H_2O$	1.060	1.154	.918
$C + O_2 \rightarrow CO_2$	1.022	1.020	1.002
150°C			
$H_2 + 1/2O_2 \rightarrow H_2O$	1.148	1.261	.910
$CH_4 + 2O_2 \rightarrow CO_2 + 2H_2O$	1.037	1.038	.999
$C + O_2 \rightarrow CO_2$	1.023	1.020	1.003

debate about the causes why such departure occurred (3). It can be
finally traced back to parasitic reactions occurring simultaneously
with the oxygen reduction reaction. In highly purified solution,
the 1.229 V has been reported (4) for this reaction under reversible conditions. Any decrease in open circuit or equilibrium
potential represents a decrease in the efficiency of the fuel cell;
in the above case ϵ goes from 0.83 at 1.229 V to 0.73 at 1.0 V.

The equilibrium potential observed is depicted in the Figure 1
where we illustrate the potential-current behavior for the cathodic
reaction (the oxygen reduction) coupled with a possible unspecified
anodic oxidation of organic impurities. At electrochemical equilibrium $i_{cathodic} = i_{anodic}$ and no net current flows in the circuit.
Since the potential for the O_2 reduction is at a fairly high positive value, the oxidation of an organic molecule is feasible and
as shown in the figure the equilibrium potential is at a value lower

than the 1.229 V expected. In this respect, we may mention the extensive work and progress made at the Defence Research Establishment by E. Criddle in which new techniques to purify the water used in electrochemical reactions have been put forward (5)

  b) Ohmic losses. The finite resistance of the electrolyte, the substrate and the membrane used in a fuel cell will induce a supplementary loss in efficiency. This reduction becomes severe at higher current densities since the power loss is proportional to the square of the current density. The solutions to this problem rely greatly on good engineering practice and on a fundamental understanding of the type of electrode used.

  c) Concentration polarization. As the load increases in a fuel cell and consequently the current increases, there may arise a situation where the rate determining step of the sequence of reactions (the slowest step) involved might be due to the finite rate of transport of the reacting species to the electrode. This situation is particularly evident for gaseous reacting species (eg. H_2/O_2 fuel cell). For planar electrodes in a solution saturated with oxygen but without stirring, a limiting current behavior will arise as shown in Figure 2. In that case the voltage of the fuel cell will decrease abruptly for any increase in the current demand decreasing accordingly the efficiency of the device.
  To reduce and overcome this concentration polarization, several techniques are available. For reactants soluble in the electrolyte, high bulk concentrations are used and also the electrolyte is circulated by pumping which uses a fraction of the electricity produced by the fuel cell and hence reduces the available power. For gaseous reactants, porous gas electrodes are used to achieve larger contact of the three phases, namely the gas, the liquid and the solid phases. There are different types of such electrode and two examples are shown in Figures 3a et 3b. In the first instance (1,3)   two types of nickel are used: on the side exposed to the gas, large pores are produced in the metal and adjacent to this structure, a network of smaller pores are produced to hold back the electrolyte. The reacting gases diffuse rapidly in the large pores and come in intimate contact with the electrolyte present in the small pores. For the electrochemical reaction to occur, a three phases contact is needed since a gaseous reactant produces a solvated reaction product and in this process an electron is given or withdrawn from a solid conducting substrate.
  In another electrode structure, a porous hydrophobic membrane is compressed on an electrode structure (3,6). The hydrophobicity of the membrane prevents the weeping of the electrolyte and allows the gas to penetrate freely into the electrode structure. This electrode consists mainly of an electroactive material (usually precious metals) dispersed on an inert substrate (typically carbon) and these components are bound together by some partially hydrophobic agent (PTFE) :the mechanical support and electrical contact is

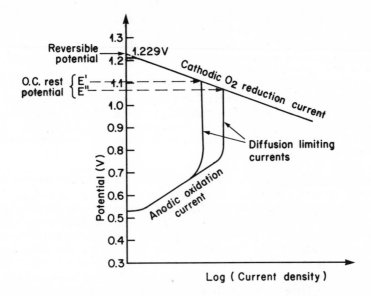

Figure 1.   Polarization curve for the electrochemical reduction of oxygen coupled with the electrochemical oxidation of an unspecified organic impurity; O.C. designates the open circuit potentials

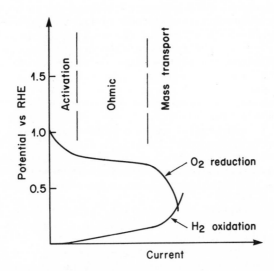

Figure 2.   Typical potential current curve for the hydrogen–oxygen fuel cell illustrating the different power losses as the current drain increases

provided by a metallic mesh or a graphite support. There has been
a major technological effort in fuel cell research to optimise the
different constraints: prevention of weeping, mechanical rigidity
and, most important, the efficient use of the electrocatalyst.
The most widely used catalyst in the acid electrolyte fuel cell is
platinum. The main effort is then to disperse as much as possible
this metal to reduce its loading without affecting adversely the
electrode performance. An additional factor to be borne in mind
in the design of fuel cell electrodes is the ease of their mass
production. Since each cell will generate at most 1 V, several
hundreds of individual electrodes must be made and assembled to
provide practical power outputs.

    d) Electrocatalytic factors. The main stumbling block to the
large scale production and use of fuel cells is the imperfect
nature of the electrocatalyst. The fundamental drawback of a fuel
cell is that it shows the best efficiency when it is not working:
in other words, as soon as we start to draw some current from it,
the useful voltage at the electrodes starts to decrease and this
decrease is linear with the logarithm of the current. The potential-current behavior of the oxygen and hydrogen half cell electrochemical reactions are shown schematically in Figure 2. At a
low current drain, the main loss is due to the activation overpotential. This loss is intrinsic to the electrocatalyst used and
it represents the extra energy needed for a reaction to proceed at
a reasonable rate. At higher current drains the ohmic losses pro-
dominate and finally at still higher current, we run into the mass
transport limitations. For a full cell based on an acid electrolyte, the oxygen reversible electrode is the most troublesome. To
evaluate the performance of an electrocatalyst three main criteria
have to be considered: the exchange current density, the Tafel
slope and finally the stability of the material in the electrolyte.
    The exchange current density, i_o, is the current per unit area
associated with the reversible reaction at equilibrium. Figure 4
represents the usual experimental technique to obtain such data
(3): the exchange current density i_o is obtained by extrapolating
the linear log i vs potential curve to the reversible potential
value. For platinum, we note the higher value for the hydrogen
reaction compared to the oxygen reaction. The main goal of electrocatalysis research is to find materials that will increase this
exchange current density to the same order of magnitude as for the
H_2 reaction on platinum.
    The second important criterion is the Tafel slope or the slope
of the linear relationship between the log i and the potential.
The lower the slope, the lower the activation loss; for the O_2
reduction reaction in acid media, the observed slope is 0.060 V/
decade and it can be obtained theoretically by assuming the oxygen
discharge step as the slowest one (the rate determining step, the
r.d.s. (7)). For hydrogen the mechanism is different and the r.d.s.
is the atomic dissociation and in this case a Tafel slope of 0.030

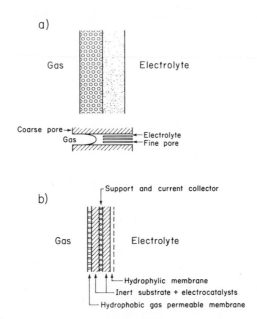

Figure 3.   Schematic of fuel-cells electrodes design: (a) metal electrode with two types of pores; (b) composite electrode with dispersed electrocatalysts

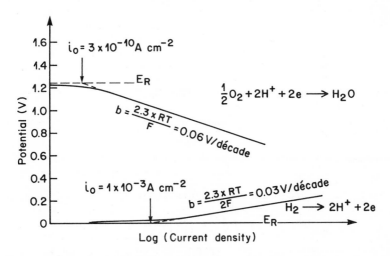

Figure 4.   Typical polarization curve of the electrooxidation of hydrogen and electroreduction of oxygen; the exchange current density, i_o, determined by extrapolation of E vs. log i to the reversible potential

V/decade is expected and found in acid electrolytes on platinum
(3).

The third aspect to consider is the electrochemical stability
of the material used. For the oxygen reduction reaction, the
electrode potential is highly anodic and at this potential, most
metals dissolve actively in acid media or form passive oxide films
that will inhibit this reaction. The oxide forming metals can form
non-conducting or semi-conducting oxide films of variable thick-
ness. In alkaline solutions, the range of metals that can be used
is broader and can include non-precious or semi-precious metals
(Ni, Ag).

A final practical criterion involves the cost of the electrode material. For terrestrial applications and for large scale
use of fuel cells, the precious metal has to be excluded or at
least limited to very small loading levels. In Figure 5, we
illustrate the quantity of Pt needed for a 600 MW generating plant
with different operating cell voltages at a given current density
of 250 ma cm^{-2}: the price tag associated with each platinum loading
is indicated for a metal price of $6.80 per gram. The present day
technology is around 1 mg or less of platinum per cm^{-2}. For 1 mg
cm^{-2} of platinum for a cell voltage of 0.7 V, the cost would be
$46.6 M; if the aims of $300 per kW (8) for a stationary fuel
cell can be attained, the platinum cost for such an installation
would represent 25% ($180 M for the 600 MW power plant). We see
that a 10 mg cm^{-2} loading would be unacceptable. Besides the
problem of price, the question of availability would cause a serious challenge to the wide-spread use of fuel cells with platinum
as the unique electrocatalyst. In the above example, the quantity
of platinum needed would be 6.8 x 10^3 kg or 0.2 x 10^6 ounces:
the world wide production of platinum is approximatively 1 x 10^6
ounces (3) per annum. As we see the number of large power system
that could be installed would be very limited. The replacement of
the precious metal in the electrode structure by another suitable
material is the most difficult challenge the electrochemical technology faces to achieve a significant breakthrough in fuel cell.
The practical down to earth avenue chosen by the industry is to
use the minimum quantity possible of platinum by suitable disper-
sive techniques (9).

III. Types of Fuel Cells

Following the extensive review of Kordesch (1), we do not need
to elaborate at great length on this subject. In Table III we list
the main systems that have been used. The candidates that are most
likely to undergo a sustained development are the hydrogen-air
fuel cells. The hydrogen fuel will be either from an hydrocarbon
feed-stock and will contain carbon dioxide and traces of carbon
monoxide. For such fuels, alkaline fuel cells are excluded. In
the event of the availability of pure electrolytic hydrogen as
proposed in the Hydrogen Economy scheme (10), alkaline electrolyte
fuel cell would then be compatible. Such fuel cells would also

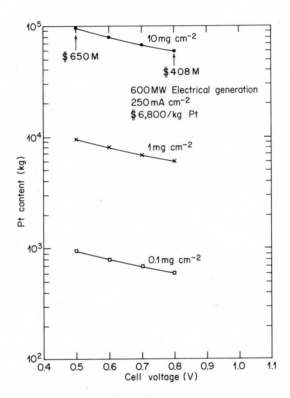

Figure 5.   Total platinum needed for a 600 MW generating fuel cell for different cell voltages; three platinum loadings (mg Pt/cm²) are illustrated

TABLE III   Fuel Cells Characteristics

Type	Electrolyte	Typical Operating Temp. (°C)	Advantages	Problems
H_2-O_2 or Air	Acid (Phosphoric)	120–190	– Use of reformed hydrogen – Use of air; no CO_2 problem	–Electrocatalyst (Pt) –Corrosion – CO poisoning
	Solid Polymer (Nafion)	150	– Same as above	–Same as above
	Alkaline (KOH)	60–150	– Semi precious electro-catalyst – Less severe corrosion	–Limited to pure H_2 and O_2
	Molten carbonate	600–700	– Ni or Ag electrocatalyst – CO_2 necessary for operation	–High temperature –Stability of materials –Corrosion
	Solid Electrolyte (Stabilized zirconia)	1000 and above	– High Efficiency – No activation losses	–Stability of materials –Limited materials withstanding these temperatures
Hydrazine-Air	Alkaline	60	– Good efficiency – Liquid fuel	–High cost of N_2H_4
Methanol-Air	Acid	60	– Liquid fuel	–High Platinum loading
Hydrocarbons-Air	Acid	60 and more	– Liquid fuel	–High platinum loading

be adequate for some Canadian purpose namely providing electrical power in remote areas from hydrogen generated in situ by electrolysis from wind power. The alkaline fuel cells are also considered in relation to the electrochemical synthesis of chlorine; in this industrial process, hydrogen is a by product and is presently burned to provide heat for the purification and drying of sodium hydroxide. Alternatively, this hydrogen could be used to provide electrical power for the production of chlorine (11).

IV. Present R and D Efforts

i) Moderate temperature fuel cells. The most important effort in dollars and man-years in fuel cell R and D has been carried out at the United Technologies Corp. (formerly Pratt and Whitney or United Aircraft), Conn., U.S.A. In the late sixties, they were involved with a series of natural gas companies in the TARGET (team to advance research for gas energy tranformation) program aimed at the development of a 12.5 kW (nominal) power stack meant to be operated on natural gas. The schematic layout of such a generating plant is shown in Figure 6 and the photograph of the actual unit is shown in Figure 7. Several such prototypes were tested by a whole range of gas and electric utilities in the U.S., Japan and Canada. Hydro-Québec (12) ran a one year test program on six such units to provide a nominal power of 75 kW. The test was performed in 1972 in Québec City. The main interest of Hydro-Québec in fuel cells is their possible use in remote areas not linked to the main power grid. Due to their intrinsic high efficiency, they would compete favourably with diesel engines now being used to generate electricity in these remote localities. Figure 7 is an actual photo of the installation used in the Hydro-Québec fuel cell test program. The fuel used was propane gas since natural gas with very low sulfur content could not be found in Québec City. The test was aimed at evaluating the response time, the coupling behavior with the network, the ease of operation and the efficiency of these new power sources. For most goals, (coupling with network, ease of operation and response time) the fuel cell fulfilled the expectations. However the lifetime of the unit was not adequate and this short lifetime was due to the inability of the system to efficiently eliminate the heat produced by the different power losses discussed previously (activation and ohmic). The overall low efficiency (below 30%) could be partly attributed to the use of propane gas instead of the natural gas, a substitution that reduces the stated performance of the reformer. The electrochemical stack itself did meet the expectations with respect to the efficiency but as stated above had too low a lifetime. The inverter presented no particular problems.

Following the TARGET program, United Technologies Corp. is developing a 40 kW unit capable of producing electricity as well as providing steam at 15-60 psig. and hot water at 41°C. With such a system between 70-90% of the energy value of the fuel can be extracted for useful purposes. The main application of such units

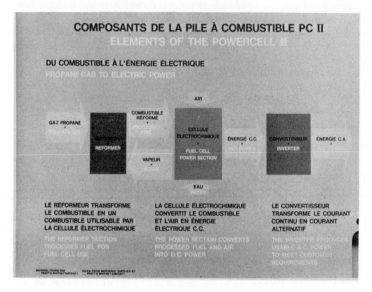

Figure 6. *Flow sheet of the 12.5 kW United Technologies fuel cell tested by Hydro–Québec*

Figure 7. *Photograph of the actual installation in Québec City (1972)*

would be to provide useful energy to apartment buildings or small
commercial centers. Figure 8 illustrates the schematic flow sheet
of such a unit (13).

More recently, a major R and D effort has been launched to
develop a 27 MW fuel cell power plant. This effort is carried out
at United Technologies in close collaboration with several electric
utilities. Presently a 1 MW power plant is under test and a 4.8 MW
demonstration unit should be installed in 1979-80. This demonstration program is sponsored by EPRI (Electric Power Research Institute) and DOE (Department of Energy). The flow sheet of this unit
is shown in Figure 9 (14). This program, the largest R & D effort
in fuel cells today, aims at producing in the near future a commercial product. The estimated cost should be in the neighbourhood
of $200-350/kW in 1975 dollars (8). The cost figures are very
difficult to assess because some R & D is still being carried out
to optimize different components and also to test the lifetime of
the different units. A 5 year lifetime for the electrochemical
stacks, 10-15 years for the fuel conditioner and 20 years for the
power conditioner are the present aims. Only the next years can
tell us if these goals will actually be met and if the fuel cells
will achieve the high hopes placed in them.

To our knowledge, there are no other such large scale R & D
efforts on fuel cells elsewhere for civilian purpose. The other
present work is scattered in different countries and is aimed at
much lower power output (10-20 kW). In Belgium, a conglomerate of
several companies established a new company, Elenco, that is now
in the process of developing some power plants based on alkaline
fuel cells (11). Some Japanese R and D effort is also being devoted to this area in the framework of the Sunshine program (15).

ii) <u>High temperature systems</u>. United Technologies has already launched a program to develop the second generation fuel cell
based on the molten carbonate system. The Institute of Gas Technology in Chicago and General Electric in Schenectady are also involved in this area of research. Their aim is to have some demonstration units by 1985. In this system, which operates at
650-700°C, no noble metals are used. This development is being
stimulated by the U.S. effort to exploit on a large scale their
coal deposits.

iii) <u>Canadian contributions</u>. As elsewhere in the world, the
Canadian R and D work directly on the fuel cells experienced its
peak activity in the late sixties-early seventies. In recent years,
it is fair to say that all the industrial efforts directly aimed
at the production or testing of fuel cells have ceased. However a
few governmental or para-governmental agencies and universities
have carried out some modest activities to either elucidate fundamental problems in the field of electrocatalysis or to evaluate
some available prototypes. Outside the universities, two main
groups of electrochemists have been focussing their efforts in this

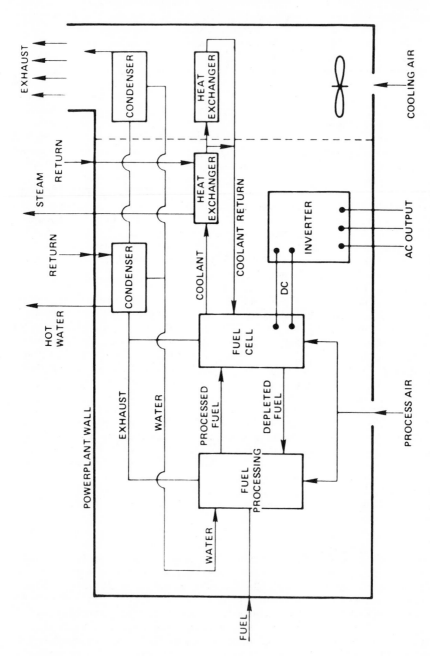

Figure 8. Flow sheet of the 40-kW fuel cell under development at United Technologies

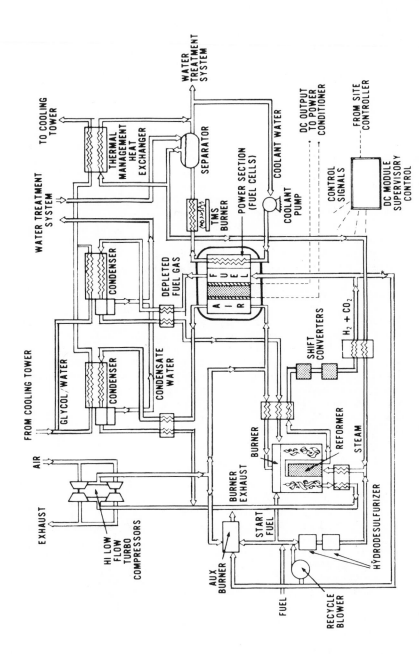

Figure 9. Flow sheet of the 4.8-MW demonstration unit being built by United Technologies for EPRI and DOE (U.S.A.)

field of research : the Defence Research Establishment in Ottawa
(Shirley Bay Laboratories) and l'Institut de Recherche de l'Hydro-
Québec (IREQ, Varennes). In Ottawa, a great effort was put in
achieving the high purity criterion to evaluate the reversible
potential of the oxygen electrode (5). The main development was
centered in the purification of the water used with the electrolyte;a patented method and instrument were established whereby the
steam from a distillation still is passed over heated platinum
when the traces of organic impurities are pyrolysed. Some development was also carried out to improve the electrode composition and
performance in the alkaline fuel cell (16). Finally a thorough
evaluation of the actual performance of a 300 W hydrazine-air fuel
cell was carried out (17).

   At IREQ, besides the participation in the field tests run by
the engineers of Hydro-Québec (12), the main effort has been to
tackle fundamental problems in the field of electrocatalysis
(18-22) and of anodic oxidation of different potential fuels (23-
26). A careful and extensive study of the electrochemical properties of the tungsten bronze has been carried out (18-20); the reported activity of these materials in acid media for the oxygen
reduction could not be reproduced and this claim by other workers
has been traced back to some platinum impurities in the electrodes.
Some novel techniques in the area of electrode preparation are also
under study (21,22): the metallic deposition of certain metals on
oriented graphite show some interesting catalytic features for the
oxygen reduction and also for the oxygen evolution reaction.

   Some work on the electrooxidation of the methanol and hydrazine has been published (23,26). For methanol (24), the electrooxidation was studied in presence of different anions to evaluate
the poisoning effect of several of them. Also, some work on anhydrous methanol was aimed at showing the role of water in the
reaction (23).

   Many fundamental research projectsand thesis bearing on the
general problems in fuel cells have been carried out in Canadian
Universities.

V.   Conclusions

   To conclude we may speculate on the future of the fuel cell
research and make some observations. After much hope of achieving
a cheap, reliable and efficient power conversion system, the hard
facts of life have shown that the rapid commercialization and wide
spread use of fuel cells cannot be obtained so readily. The level
of R and D funding after the bonanza years of the moon flight
missions rapidly decreased in the late sixties and several programs
in industry came to a halt. The recent and exhaustive review by
Kordesch (1). on fuel cell development indicates, among other
things, that the peak in R and D in this field was in 1964. The
maximum number of publications in fuel cell research activity occurred in 1969.

   Since 1973, the emphasis on the energy-related research in
industrialized countries has rekindled the interest in fuel cells.

In a relatively recent symposium organized by EPRI on fuel cell research (9), the pertinent problems and solutions foreseen were reviewed. The hopes to find new electrocatalysts for the oxygen reduction in acid electrolyte were concluded to be rather dim. The main effort now is to devise the best format in which platinum or other precious metals will be the most effective. The optimum reduction in crystallite size has to be determined and, especially, the lifetime of such active surfaces. The problem of CO poisoning found in the use of reformed gas is still a major source of con- cern: the use of relatively high temperatures (190°C) is the most straight forward solution to this problem but it induces severe material deterioration. Much fundamental research is still needed to achieve a clearer understanding of the basic questions of the electrode mechanisms, the exact role of the platinum oxides, the inhibition effect of CO and sulfur containing compounds. The chemical engineering espects of the power stack is of cruciaJ im- portance: adequate flow and distribution of reactants, electrode design and, most important, the excess-heat removal to avoid over- heating problems. Important effort is still needed to achieve the final goal of efficient, long lifetime, competitive and trouble- free fuel cells.

Acknowledgements

The author would like to express his gratitude to several people who have contributed to the elaboration of this paper by providing technical information: Dr. E.J. Casey and E. Criddle, Defence Research Establishment, Ottawa, Ont., W. Lueckel, Jr., United Technologies, South Windsor, Connecticut, U.S.A. and M.H. Van den Broeck, Elenco, Mol, Belgium. The help and guidance of Dr. A.K. Vijh of IREQ is also acknowledged.

Abstract

The fuel cell is an energy conversion device that theoreti- cally can convert directly the free energy of a chemical reaction into electrical power. As all energy conversion systems, the fuel cell have theoretical limits expressed as the thermodynamic effi- ciency defined as the ratio of the change of the free energy (ΔG) on the change in enthalpy (ΔH). This theoretical efficiency is, for most reactions, very high (over 85%). However, the actual fuel cells are most of the time much lower in efficiency: several factors lower the conversion efficiency, namely the activation and concentration polarizations and the ohmic losses are among the most important.

After an intensive period of R and D on fuel cells in the fifties and early sixties, the present day efforts are centered around one major development under way at United Technologies Corp. (U.S.) where a 1 MW stationary testing unit is now completed and a 4.8 MW demonstration system is under elaboration. Finally, a 27 MW stationary electrical power generating system is due for the late eighties. The present day Canadian effort is minimal and is centered on fundamental aspects linked to energy conversion devices and in the evaluation of available units.

Literature Cited

1. Kordesch, K.V., J. Electrochem. Soc., (1978),125, 77C.

2. Grove, W.R., Phil. Mag., (1839),14, 127.

3. Bockris, J. O'M. and Srinivasan, S., in "Fuel Cells: Their Electrochemistry", Mc Graw-Hill, New York, 1969.

4. Bockris, J.O'M. and Hug, A.K.M.S., Proc. Roy. Soc., Ser. A (1956),237, 277.

5. Criddle, E.E., "Proc. Symp. Oxide Electrolyte Interface", R.S. Alwitt, (Ed.), The Electrochem. Soc., Princeton, 1973.

6. Niedrach, L.W. and Alford, H.R., J. Electrochem. Soc., (1965), 112, 117.

7. Damjanovic, A. and Brusic, V., Electrochim. Acta, (1967),12, 615.

8. Fickett, A., EPRI Journal, (1976),April.

9. Fuel Cell Catalysis Workship, Special Report Electric Power Res. Inst., Palo Alto, 1975.

10. Gregory, D.P., in "Electrochemistry of Cleaner Environments", J.O.'M. Bockris, (Ed.), Plenum Press, New York, 1972.

11. Van Den Broeck, H., Elenco Company, private communication.

12. Théorêt, A., 89th Annual Meeting of the Engineering Inst. of Canada, Winnipeg, October, 1975.

13. Bowlan, P. and Hanoley, L.M., in "First Generation Fuel Cell Powerplant Characteristics,"United Technol. publication, Nov. 1977.

14. Handley, L.M., Rogers, L.J. and Gillis, E., Proc. 12th Inter. Soc. Energy Conv. Eng. Conf., (1977),34.

15. Tamura, K., Hitachi, Ltd., private communication

16. Armstrong, W.A., U.S. patent 4,020,239 (1975), Can. Patent 1,016,600 (1975)

17. Gardner, C. Fouchard, D., Sawchuk, R. and Hayashi, R., Ontario-Québec Electrochemical Society Meeting, Toronto, Feb. 1977.

18. Randin, J.P., J. Electroanal. Chem., (1974),51, 471.

19. Randin, J.P., Can. J. Chem., (1974),52, 2542.

20. Randin, J.P., J. Electrochem. Soc., (1974), 121, 1029.

21. Morcos, I., J. Electrochem. Soc., (1975),122,1008.

22. Morcos, I., J. Electrochem. Soc., (1975),122, 1492.

23. Bélanger, G.,J. Electrochem. Soc., (1976),123, 818.

24. Bélanger, G., Can. J. Chem., (1972),50, 1891.
25. Bélanger, G. and Vijh, A.K., Surf. Technol., (1977),5, 81.
26. Vijh, A.K., J. Catalysis, (1974),37, 410.

RECEIVED September 25, 1978.

19

Chemical Trends in the Nuclear Power Industry

S. R. HATCHER

Atomic Energy of Canada Limited, Ottawa, Canada

Although the nuclear power industry clearly involves physics, mechanical and electrical engineering, it is not self-evident that chemistry would be important. However, a brief study shows that chemists and chemical engineers play a vital role, from the production of essential materials such as uranium and heavy water for CANDU (CANada Deuterium Uranium) reactors, through the design and operation of nuclear generating stations, to the management of radioactive materials remaining after power generation.

I think it might be useful to first review some of the more important chemistry challenges and achievements in the development of the nuclear power industry up to the present time. Then I will indicate where the chemical research and development effort will be most active in the future and what the needs of the industry may be for chemists and chemical engineers.

Development of the Industry

The essential ingredients for producing heat in a thermal fission nuclear reactor are the fuel and a moderator. A heat transport system with its coolant is necessary to convey the heat from the reactor to boilers where steam is produced to drive the turbogenerator. The natural materials available for fuel and moderator are uranium ore and water; natural uranium extracted from the ore comprises the fissionable isotope uranium-235 and water contains hydrogen which is a good moderator. (Table I)

TABLE I

Nuclear Materials

	Fuel	Moderator
Material Form	Uranium Dioxide	Water
Essential Element	Uranium	Hydrogen
Desirable Isotope	^{235}U	^{2}H (D)
Natural Occurrence in Element	7×10^{-3}	1.5×10^{-4}

But nature provides only 0.7% of the fissionable isotope ^{235}U in natural uranium, the rest being ^{238}U; and although the predominant hydrogen isotope ^{1}H in water is a good moderator, it absorbs too many neutrons to allow a reaction to be maintained with the low ^{235}U content of natural uranium. However, the heavy hydrogen isotope ^{2}H or deuterium, which occurs at a concentration of about 150 parts per million, is an excellent moderator because it absorbs very few neutrons. So to obtain a practical combination of uranium fuel and water moderator, either the uranium must be enriched in ^{235}U for fuel or heavy water must be used as the moderator.

Because early Canadian reactors used heavy water, and because it is also fundamentally the most efficient moderator, Canada naturally adopted the heavy water reactor for the development of a nuclear power system. By using heavy water both as moderator and as coolant, and by refuelling with the reactor at power, it was possible to develop the CANDU system to operate efficiently and economically with natural uranium fuel. This in turn resulted in the simplest possible fuel cycle.

Fuel. The nuclear fuel cycle starts with mining of the uranium ore, chemical leaching to extract the uranium, and solvent extraction with tributyl phosphate to produce eventually pure uranium oxide. If enriched uranium is required, the uranium is converted to the gaseous uranium hexafluoride for enrichment by gaseous diffusion or gas centrifuge techniques, after which it is reconverted to uranium oxide. Since the CANDU system uses natural uranium, I will say no more about uranium enrichment although, as I'm sure you appreciate, it is a major chemical industry in its own right.

Uranium dioxide for use in nuclear fuel must be produced to a stringent specification so that it can be pressed into pellets and sintered at high temperature in hydrogen to produce dimensionally stable, crack-free UO_2 pellets with a density typically 97% of theoretical. The fuel pellets are loaded into zirconium alloy tubes, welded closed and assembled into fuel bundles.

During operation in the reactor, many fission product elements are formed in the UO_2 fuel. Of particular importance in fuel design are the fission product gases such as xenon and krypton, which can be released from the UO_2 crystal structure at high temperatures. Free fission gas can increase the pressure in the fuel element and lead to operating problems if a fuel element should develop a defect. Thus a thorough understanding of the chemistry and physics of the entire pellet production process had to be developed (1), together with a detailed knowledge of the mechanisms governing the behavior of the noble gases and other fission products in UO_2 at temperatures up to 2600°C. (2)

Zirconium alloys are used as in-reactor materials because of their very low neutron absorption, high strength and excellent corrosion resistance. However, they do absorb hydrogen freely.

As the solubility is exceeded in the zirconium, precipitation of
zirconium hydride commences and ultimately the ductility of the
alloy can be reduced leading to the possibility of cracks. Thus
an additional requirement in fuel development was a thorough
understanding of the hydriding mechanism, sources of hydrogen,
rate controlling steps, protective methods, specifications of
materials and processes, and quality assurance to achieve the required performance. (3)

Nuclear fuel production is now a maturing industry in Canada
with two competing industrial companies well established and a
third about to enter the field. The performance of Canadian fuel
has been excellent with a remarkably low defect rate of 0.03%
since 1972. (4) Further process improvements are being made con-
tinually as a result of efforts to be more cost-effective while
maintaining the necessary high standards.

Heavy Water. Heavy water is produced by an initial isotopic
separation to extract and concentrate the naturally occurring
deuterium isotope from hydrogenous material. The most abundant
source of starting material is water, although processes have been
developed for hydrogen and natural gas. The only process in commercial use today is the Girdler Sulfide (GS) process which uses
a bithermal chemical exchange between hydrogen sulfide and water
to concentrate from the natural level of about 150 $\mu gD/gH$ to 20-
30%. Water distillation is used to achieve the final product
concentration of 99.7% D_2O.

The GS enriching process is a counter-current gas-liquid extraction done at a pressure of 2000 kPa in a sieve tray tower
with the upper half operating at $30^{\circ}C$ and the lower at $130^{\circ}C$. (5)
In the top half of the tower, feedwater extracts deuterium from
the upflowing cold H_2S, reaching a maximum at the centre of the
tower. The recycled lean H_2S entering the lower hot half of the
tower strips deuterium from the water, which then leaves the
system depleted in deuterium. A cascade of several stages is
used to reach the desired feed concentration for the final water
distillation or finishing unit. Transfer between cascades can be
either by gas or liquid from the centre of the tower.

The process has been used in the USA for 25 years, but a
large extrapolation of experience was necessary to reach the scale
of operation needed for the Canadian nuclear industry. A typical
Canadian enriching unit has a nominal capacity of 50 kg D_2O/h.
However, with an overall extraction of about 20% of the deuterium
in the feedwater, the total feed rate is about 0.5 metric tons
per second. Three large towers are used in parallel. (Table II)

These towers are amongst the world's largest high pressure
chemical process units and their successful performance in this
service required development of a new area of operational experience.

TABLE II

Typical First Stage Canadian GS Enriching Units

Tower Diameter, m	- 8.6
Tower Height, m	- 90
Sieve Trays per Tower	- 130
Cold Temperature, ^{o}C	- 30
Hot Temperature, ^{o}C	- 130
Pressure, kPa	- 2000
H_2S Flow per Tower (Recycled), kg/h	- 20×10^5
Feed Water Flow per Tower, kg/h	- 6×10^5
Feed Water Deuterium, Fraction	- 150×10^{-6}
Discharge Water Deuterium, Fraction	- 120×10^{-6}
Number of Towers	- 3
Total First Stage D_2O Production, kg/h	- 50

Control of the liquid:gas (L/G) ratio in the towers is critical in maintaining the design extraction performance; relatively
small changes can result in significant loss of extraction.
Startup experience indicated that at design conditions foaming on
the trays resulted in instabilities which upset the L/G ratio and
caused loss of production. A large R&D effort was mounted, both
in the laboratories and at the plants, to understand the controlling mechanisms and to develop design and operational techniques
to overcome the problem. The key feature proved to be surface effects at the gas/liquid interface at pressures near the H_2S liquefaction point. Good performance was achieved by a combination of
design changes to the trays, chemical control and the accumulation
of operating experience to recognize the symptoms of incipient
instability and to take remedial action.

Another phenomenon which proved to be of operational concern
was the deposition of iron pyrite, FeS_2, which I will discuss
later.

Fuel Recycle and Radioactive Waste Management. In the 1950's
all countries developing nuclear power were examining the reprocessing of fuel to recover useful fuel materials for recycle. The
processes studied involved dissolution of the fuel in nitric acid
followed by liquid-liquid extraction using an organic solvent such
as tributyl phosphate diluted with kerosene. In a multi-stage
counter-current process, the useful materials, plutonium and uranium, were co-extracted into the organic phase while the fission
products were rejected as waste in the aqueous raffinate. Then by
chemical reduction of the plutonium in the solvent, a partition
could be made between plutonium and uranium. Finally, each could
be purified by further solvent extraction or ion exchange.

In parallel Canada also pioneered work on the immobilization
of radioactive wastes into glass (vitrification) for permanent
disposal. (6) A natural mineral, nepheline syenite, was used as
the basic material because it produced a glass with excellent

leach resistance. The aqueous acidic fission product waste was
fed into a crucible together with the nepheline syenite and a
flux. The acid and moisture were driven off and the residue
dried and fused to give a stable, leach-resistant product. The
technical feasibility of immobilization for waste disposal was
thus established. A disposal experiment was initiated with 25
glass blocks containing fission products buried under adverse
conditions in wet sandy soil at Chalk River to measure the leaching of the glass and movement of radionuclides with time. (7)

Meanwhile, success in the development of the natural uranium
fuelled CANDU concept had led to very low cost fuelling and effective utilization of uranium even without recovery through reprocessing. AECL therefore decided to set aside work on reprocessing and concentrate instead on the once-through fuel cycle with
storage of the irradiated fuel. The evidence indicated that the
zirconium clad UO_2 fuel could be stored under water for many decades until a decision was needed regarding recycle or disposal.

Mass Transport at Very Low Concentrations. Reactor Circuits.
Early in the development of water-cooled reactors, it became
apparent that at temperatures of $250-300^{\circ}C$ with a non-isothermal
circuit, corrosion of carbon steel could lead to significant mass
transport of iron if the chemistry of the system were not properly
controlled. The resulting buildup of large deposits of "crud" on
fuel surfaces caused fuel failure. However, the large cost differential between carbon steel and stainless steel provided an
incentive to identify chemistry conditions for the successful use
of carbon steel.

Little was known of the thermodynamics of metal, water, oxygen and hydrogen systems at these temperatures or of the kinetics
of reactions involved. Additional R&D quickly established that
the basic requirements were for a pH of about 10 and an absence of
oxygen. Lithium hydroxide was used for pH control to avoid neutron absorption and radioactivity buildup with sodium or potassium.
The intense radiation field in the core of a reactor results in
radiolytic reactions with a net decomposition of the water to hydrogen and oxygen. Thus it became necessary to provide an excess
of hydrogen to displace the reactions away from the natural steady
state involving free oxygen. This chemical control of high pH and
excess hydrogen led to excellent fuel performance.

After a few years of operation there was a significant increase in radiation fields from the primary circuit piping in the
Douglas Point generating station. Other water-cooled reactors
around the world experienced similar effects. The principal
source of the radioactivity was traced to cobalt-60, formed by
neutron absorption in the natural cobalt-59 which arose from hard-
facing alloys and was also present as an impurity in boiler materials such as monel, and in carbon steel and other structural
materials. The mechanism of this radioactivity transport was
found to be corrosion of the cobalt bearing materials, transport

of the corrosion products into the reactor, deposition on the
reactor fuel surfaces where neutron activation takes place, followed by another transport step and deposition of the activated
material onto piping outside the reactor. The actual quantities
of material involved in the process are extremely small under
good conditions such as in Pickering; typically about 0.1 g Fe/m^2
deposited on the fuel and 2×10^{-4} g Co/m^2. Thus over the entire
core area of 3900 m^2, the total deposition is about 390 g Fe and
0.8 g Co. (8) The process has been modelled mathematically. (9)
Activity transport can be minimized by eliminating cobalt bearing
materials and ensuring rigid adherence to chemical specifications
for the coolant. Various decontaminating procedures have been
used including on-line chemical and thermal cycling and the use
of proprietary dilute decontaminating procedures such as CANDECON.
(10) The radiation fields in Douglas Point were reduced markedly
and excellent control of activity transport is being achieved in
Pickering.

Mass Transport at Very Low Concentrations. Heavy Water
Plants. The phenomenon of mass transport at very low concentrations is not unique to the reactor coolant systems. It can occur
also in the heavy water production plants. Table III compares
iron transport in a reactor primary circuit and a GS plant dehumidifier circuit and illustrates the quantities of iron that can be
transported each day. While the concentrations in the reactors
are typically two orders of magnitude lower, the flow rates are
an order of magnitude higher. The lower concentration in the
reactors gives a lower driving force for deposition and the efficiency of deposition is considerably lower.

TABLE III

Mass Transport at Very Low Concentrations

	Pickering Reactor Primary Circuit	BHWP Steam Heater (early operation)
Fluid Flow, kg/s	7800	400
Approximate Concentration, μg Fe/kg	10	1000
Fe Transported, kg Fe/day	7	35
Approximate Deposition Rate, kg Fe/day	0.02	10
kg Co/day	3×10^{-5}	

In the first years of operation of the Bruce Heavy Water
Plant (BHWP) deposition of iron pyrite, FeS_2, necessitated frequent cleaning of the dehumidifier steam heaters. The iron

transport problem has gradually decreased in BHWP due partly to replacement of carbon steel pipe by stainless steel in some critical areas of the dehumidifier loop. A similar problem of FeS_2 deposition in the holes of the hot tower sieve trays has also been overcome by better chemical control and increased gas velocities.

The importance of chemistry in other reactor circuits will be described in the next two papers by Balakrishnan and Lister (11) and Shoesmith (12).

Future Trends

The importance of chemistry to the nuclear power industry is now well recognized. Chemical control in water circuits is an accepted part of the operating requirements of nuclear generating stations, as it is for modern fossil-fired stations. While there have been major advances in knowledge of the chemistry of aqueous systems at temperatures above 100°C, there is still a need for further work. As we improve our understanding of thermodynamics and kinetics of solid-aqueous reactions and the effect of radiation on them, we can expect further advances in controlling radiation fields in reactor circuits and in minimizing iron deposition in GS plants.

In the heavy water production plants a brief examination of the temperature change and water flowrates reveals the enormous amount of energy required and the importance of efficient and economic heat exchange. In fact the cost of heavy water is dominated by two components, energy and capital. Consequently there is a large incentive to implement any advances which result in an increase in extraction efficiency or total throughput. Such advances are under continual investigation and being adopted as their effectiveness is demonstrated. Further optimizations are likely for some time to come. Meanwhile new processes have reached the point where the next step in their development would be demonstration scale work (13). Of particular interest is a hydrogen amine exchange process which holds promise of economic viability at low throughputs relative to GS plants. A convergence of the availability of large hydrogen streams and the need for new heavy water production capacity may offer an opportunity for the introduction of such processes.

Now that the first generation of nuclear power plants is well established, more attention is being focussed on the R&D required to ensure that nuclear power can continue to contribute to the energy supply for the foreseeable future. Two areas which are interdisciplinary but involve a large chemical input are waste disposal and fuel recycle.

Radioactive Waste Disposal. There are two principal types of radioactive materials produced in the operation of nuclear generating stations. Over 99% of the radioactivity produced is

contained within the fuel bundles which are discharged at the end
of their useful lives. The volume of these is relatively small
but they require special handling because of their high levels of
radioactivity and the consequent release of heat as the radionuclides decay to stable nuclides. Reactor waste is the second type,
comprising materials which become contaminated during the routine
operation of the reactor. These include ion exchange resins,
filters, rags, paper, etc. Their radioactivity level is very much
lower but their volume can be substantially higher. Table IV
shows a comparison for a station of 1000 MWe.

TABLE IV

Annual Production of Radioactive Materials
(1000 MWe)

	Radioactivity, Ci (1 year cooled)	Volume, m^3
1. Irradiated Fuel	10^8	70
2. Reactor Waste		
Ion Exchange Resins	1500	25
Filters	500	5
Combustible	5	350
Liquids	5	1000

Irradiated Fuel. Although much of the plutonium produced in
CANDU reactors is consumed to produce energy while the fuel is
still in the reactor, the residual energy content of the discharged fuel is still at least as great as an equivalent amount
of fresh uranium. However, to obtain that energy would require
recovery and recycle of the plutonium; this is not economic today
nor likely to be for many years. To put the quantities of energy
into perspective, one can observe that the energy content of irradiated Canadian fuel generated in this century is likely to be
equivalent to several billion barrels of oil, comparable to our
present reserves of conventional oil.

Thus the consistent Canadian philosophy has been to store
irradiated fuel retrievably until a decision on its ultimate dis-
position is necessary or desirable. Experience with the storage
of CANDU fuel now extends over 15 years and has provided confidence that this type of fuel can be stored safely and economically
for the order of five decades, using proven technology.

In storage there is a clear intent to retrieve and further
handle the material at some time in the future. Disposal is taken
to mean that there is no intent to retrieve the material, and that
in the long term there will be no need for further human intervention.

In common with the programs of other countries, AECL intends
to develop and demonstrate the technology for the disposal of
radioactive waste produced in the operation of nuclear electric
generating stations. Two basic options are available as shown in
Figure 1. Following storage either at the reactor sites or at a
central storage facility one may:

a) dispose of the fuel, or
b) extract useful materials and dispose of the waste.

Both of these routes require immobilization of the material
to ensure a low rate of solution by water; both types of immobilization involve a large chemistry component. The early approach of
vitrification pioneered by AECL at Chalk River some 20 years ago
has stood the test of time (7) and has now been adopted and further developed internationally for handling separated waste. A
variety of techniques will be explored for the immobilization of
fuel.

Prospects for the disposal of the immobilized materials have
been examined by several groups of experts around the world
(14, 15, 16, 17, 18) and by a group headed by Dr. F.K. Hare,
Director of the Institute of Environmental Studies, University of
Toronto (19). An international consensus has emerged on the
safety and suitability of disposing of immobilized radioactive
waste by emplacement deep underground (300-1000 m) in a variety
of geologic strata.

The disposal facility in Canada is expected to be such an
engineered repository. After reviewing the requirements for a
radioactive waste repository and work done in other countries,
geologists of the Geological Survey of Canada (GSC) recommended
that intrusive igneous rock structures and salt deposits should
be evaluated as potential host rocks for the repository. Although
both types of formation are being evaluated, the emphasis is being
placed on igneous rock.

A detailed discussion of the chemical aspects of this radioactive waste disposal program is given by Tomlinson at this symposium (20).

Reactor Wastes. The same fundamental approach of immobilization and disposal is being taken for reactor wastes. Work has
been underway in AECL for several years on immobilization techniques (21). These include volume reduction processes of incine-
ration for combustible materials and reverse osmosis for concentrating solids from aqueous streams. The concentrates from these
processes will be immobilized in bitumen. The deep underground
disposal facility developed for fuel wastes will most likely also
be used for the immobilized reactor wastes.

Fuel Recycle. Although the commercial CANDU reactors use the
once-through natural uranium fuel cycle, it has been recognized
for many years (22) that exceptional uranium utilization could be

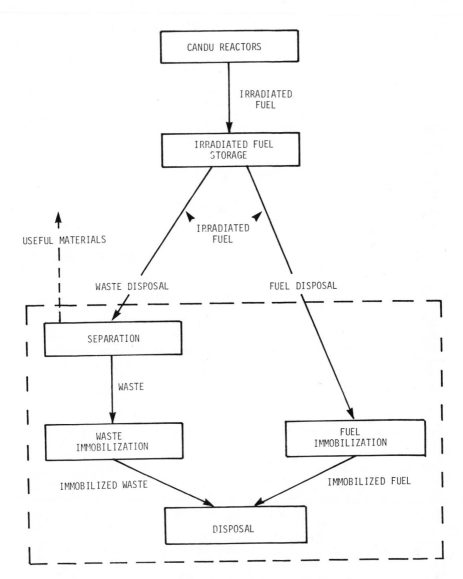

Figure 1.   Management of irradiated Candu fuel

achieved by the use of thorium in CANDU reactors. In a review of
energy sources, Hart (23) has indicated the enormous energy potential of Canada's estimated uranium and thorium resources compared to all fossil fuels (Table V). The values for uranium and
thorium fission assume efficient recycle of fissionable materials.

TABLE V

Reasonably Proven Canadian Energy Resources
from Hart (23)

Resource	Energy Content(Q)	Reasonably Proven(Q)
Coal and Lignite	2.4	0.2
Total Conventional Oil	0.4 - 0.6	0.047
Oil Sands	1.4	–
Alberta Heavy Oil	0.15	–
Total Conventional Gas	0.8 - 1.2	0.069
Nuclear Fission	$6 \times 10^4 - 6 \times 10^6$	66
Nuclear Fusion	$6 \times 10^4 - 6 \times 10^6$	66

$$1 \ Q \ = \ 10^{18} \ BTU$$
$$= \ 1.05 \times 10^{21} \ Joules$$

Current Canadian Consumption $\simeq$ 0.008 Q/a

A major advantage of the Canadian system is that the basic
CANDU reactor design can be used for efficient thorium fuel
cycles. Development of a new reactor, such as the fast breeder
reactor, is not necessary. Thorium fuelling could be initiated
using either ^{235}U or plutonium as fissionable material. Of particular interest is the fact that a 30 year accumulation of spent
fuel (4130 MgU) from a 1000 MWe natural uranium fuelled CANDU
reactor could provide sufficient material to start up and operate
1400 MWe for 30 years on a high burnup thorium cycle. Alternatively, it could be used to start up 2200 MWe and operate inde-
finitely on a lower burnup self-sufficient thorium cycle; this
would have a higher fuel cycle cost due to the more frequent recycling necessitated by the lower burnup. Thus the stocks of
irradiated natural uranium in retrievable storage offer excellent
fuel cycle flexibility for the CANDU system.

AECL has evaluated some of the basic information and development requirements in some detail (24, 25) and has outlined the
type of fuel recycle development program which would be required.
It would involve research and development of thorium fuels and
fuel fabrication methods, reprocessing, demonstration of fuel
management techniques and physics characteristics in existing
CANDU reactors and demonstration of technology in health, safety,
environmental, security and economics aspects of fuel recycle.
The program would have to provide all the necessary information
for a decision on industrial scale implementation of fuel recycle.

It would require 20 to 25 years to complete the research, development and demonstration program. This is compatible with the present Canadian uranium resources situation. Chemistry and chemical engineering would play a major role in most areas of such a fuel recycle program.

Meanwhile AECL and other Canadian departments and agencies are participating actively in the International Nuclear Fuel Cycle Evaluation (INFCE) to study all fuel cycle options. No decisions on expansion of the present research level on thorium fuels will be taken until information from INFCE has been evaluated by the Canadian Government.

How Many Chemists and Chemical Engineers?

There is no detailed documentation of the number of chemists and chemical engineers employed in the nuclear power industry. Within AECL there are 300 in a total staff of 6000 (5%). Within Ontario Hydro (26) there are approximately 145 in a total staff of 3300 associated with nuclear power generation (4.4%). The Canadian Nuclear Association (CNA) estimates that in 1976 there were about 18,400 people employed in the Canadian nuclear industry, excluding the uranium industry (27). If about 4% of these were chemists or chemical engineers, one can estimate that a total of about 700 were employed in the industry at that time. There is likely to be considerable expansion of the industry by 1985, particularly in the utilities such as Ontario Hydro, Hydro Québec, and New Brunswick Power which already have additional nuclear capacity under construction. The expansion will in turn provide new opportunities for members of this profession.

Summary

The nuclear electric power industry is now well established as an energy producer. Its development has produced many chemical challenges, some traditional, some unique. Chemists and chemical engineers have been highly successful in meeting these challenges and thereby have made a major contribution to the excellent performance of the CANDU nuclear system. The high standards of chemical performance must be maintained. Some areas for further advances are already identified and, like any other industry, new challenges will arise. As the industry expands in response to energy demands, chemists and chemical engineers will be called upon in increasing numbers to participate in this strategic element of the nation's long term energy supply.

"LITERATURE CITED"

1   Chalder, G.H., Bright, N.F.H., Patterson, D.L., Watson, L.C.
    "The Fabrication and Properties of Uranium Dioxide Fuel"
    AECL-602, 1958

2   Robertson, J.A.L.   "Introductory Survey on Swelling and Gas
    Release" Physical Metallurgy of Reactor Fuel Elements
    Metals Society, London, 1975

3   Ells, C.E.   "Hydride Precipitates in Zirconium Alloys"
    J.Nuc.Mat. (1968) 28, 129-151

4   Fanjoy, G.R., Bain, A.S.   "CANDU Fuel - Fifteen Years of
    Power Reactor Experience" AECL-5711, 1977

5   Lumb, P.B.   "The Canadian Heavy Water Industry"
    J.Brit.Nuc.Energy Soc. (1976) 15 (No. 1) 35

6   Watson, L.C., Aikin, A.M., Bancroft, A.R.
    "Disposal of Radioactive Wastes" p.375, IAEA Vienna
    STI/PUB/18, 1960

7   Merritt, W.F.   "The Leaching of Radioactivity from Highly
    Radioactive Glass Blocks Buried Below the Water Table -
    Fifteen Years of Results" IAEA Vienna IAEA-SM-207/98, 1976

8   Burrill, K.A.   Private communication

9   Burrill, K.A.   "Corrosion Product Transport in Water Cooled
    Nuclear Reactors - Part 1, Pressurized Water Operation"
    Can.J.Chem.Eng. (1977) 55, 54

10 LeSurf, J.E.   "Control of Radiation Exposures at CANDU
    Nuclear Power Stations" J.Brit.Nuc.Energy Soc. (1977) 16
    (No. 1) 53

11 Balakrishnan, P.V., Lister, D.H.   "The Chemistry of the Water
    Circuits in CANDU Power Reactors" Paper E26 Chemistry for
    Energy Symposium, Chemical Institute of Canada, Winnipeg,1978

12 Shoesmith, D.W.   "Chemical Processes Involved in Boiler
    Circuit Cleaning" Paper E27 Chemistry for Energy Symposium,
    Chemical Institute of Canada, Winnipeg, 1978

13 Rae, H.K.   "Selecting Heavy Water Processes" ACS Symposium
    Series No. 68, Separation of Hydrogen Isotopes, H.K. Rae
    (Editor). American Chemical Society, Washington, 1978

14 Polvani, C. (Chairman) et al   "Objectives, Concepts and
    Strategies for the Management of Radioactive Waste Arising
    from Nuclear Power Programs" OECD Nuclear Energy Agency,
    Paris, 1977

15 IAEA/NEA "Management of Radioactive Wastes from the Nuclear
    Fuel Cycle" IAEA, Vienna, 1976

16 Flowers, Sir Brian (Chairman) "Nuclear Power and the
    Environment" Sixth Report of the Royal Commission on
    Environmental Pollution, HMSO, London, 1976

17 Hebel, L.C. (Chairman) "Nuclear Fuel Cycles and Waste
    Management", Report to the American Physical Society,
    New York, 1977

18 Keeny, S.M. (Chairman) "Nuclear Power Issues and Choices"
    Report of the Ford Foundation's Nuclear Energy Policy Study
    Group, Ballinger Publishing Company, Cambridge, 1977

19 Aikin, A.M., Harrison, J.M., Hare, F.K. (Chairman)
    "The Management of Canada's Nuclear Wastes"
    Energy, Mines & Resources Canada, Ottawa, 1977

20 Tomlinson, M.T. "Chemistry for Millenia" Paper E28
    Chemistry for Energy Symposium, Chemical Institute of Canada,
    Winnipeg, 1978

21 Charlesworth, D.H., Bourns, W.T., Buckley, L.P. "The
    Canadian Development Program for Conditioning Reactor Wastes
    for Disposal" Paper 78 NE-18 at the ASME/CSME Joint Pressure
    Vessels & Piping Conference, Montreal, 1978

22 Lewis, W.B. "How Much of the Rocks and Oceans for Power?
    Exploiting the Uranium-Thorium Fission Cycle" AECL-1916, 1964

23 Hart, R.G. "Sources, Availability and Costs of Future Energy"
    AECL-5816, 1977

24 Hatcher, S.R., Banerjee, S., Lane, A.D., Tamm, H., Veeder,J.I.
    "Thorium Cycle in Heavy Water Moderated Pressure Tube (CANDU)
    Reactors" American Nuclear Society Meeting, San Francisco
    AECL-5398, 1975

25 Critoph, E. "The Thorium Fuel Cycle in Water-Moderated
    Reactor Systems" Paper IAEA-CN-36/177 at the IAEA Inter-
    national Conference on Nuclear Power and its Fuel Cycle,
    Salzburg. AECL-2705, 1977

26 Montford, B.M. Private communication

27 Canadian Nuclear Association "Nuclear Canada Yearbook 1977"
    Toronto

RECEIVED September 25, 1978.

Chemistry for Millennia

Chemistry Research Topics for Long-Term Retention of Radioactive Wastes Deep Underground

M. TOMLINSON

Atomic Energy of Canada Limited, Whiteshell Nuclear Research Establishment, Pinawa, Manitoba, Canada ROE ILO

This paper is concerned primarily with the application of chemistry to the control of radioactive waste products from the use of nuclear energy. As far as immediate effects are concerned, nuclear power from uranium is a particularly clean energy source (1). The radioactive waste products are well contained within the used fuel bundles. Since some constituents of the radioactive wastes take almost a thousand years to decay to an innocuous level and a few persist for many millennia, e.g. ^{239}Pu, we have to ensure that the wastes remain well contained for this length of time. This is to be achieved by deep underground disposal of the wastes, the objective of which is to isolate them from man and the biosphere until they become innocuous. This time span is short, geologically speaking, i.e. in comparison with the periods of time that many deep underground geologic formations are known to have remained stable.

The underground disposal system consists in essence of a series of barriers which prevent disturbance of the wastes and inhibit escape of radioactive nuclides. In general, each barrier alone and independently of the rest is capable, under appropriate circumstances, of retaining the wastes for the requisite time period. Several such barriers together can provide a high degree of assurance that adequate isolation is maintained. The basic components of the system are:

- *The waste form* itself which will consist of a stable solid matrix of low solubility.
- *Metal or ceramic containers* around the waste form which may be intended for long-term or very short-term containment.
- *An excavated zone which has been backfilled and sealed.* This will influence the flow of groundwater to and from the wastes and the chemical composition of the groundwater. Also It will adsorb water-borne radioactive nuclides and retard their movement.
- *A geologic formation, consisting of several hundred metres of rock and overburden,* which performs several functions. It

prevents weathering, erosion of the wastes and disturbance by plants, animals and the activities of man. Also, it inhibits the flow of groundwater to and from the wastes, and controls the chemical composition of the groundwater. Moreover, adsorption on the rock will retard the movement of water-borne radionuclides.

The task of the chemist is to delineate the conditions under which the chemical functions of the barriers will be performed satisfactorily. In this paper, I wish to review what we need to know about the chemical behaviour of this system in pragmatic terms and then attempt to summarize what advances in fundamental chemical knowledge and data will help to answer these questions.

Since transport by water is virtually the only available mechanism for escape, we will be predominantly concerned with the chemistry of aqueous solutions at the interface with inorganic solids - mainly oxides. These will be at ordinary to somewhat elevated temperatures, 20-200°C, because of the heating effects of radioactive decay during the first millennium. The elements primarily of interest (Table I) are the more persistent fission products which occur in various parts of the periodic table, and the actinides, particularly uranium and thorium and, most important of all, plutonium.

TABLE I

SOME IMPORTANT RADIONUCLIDES

^{14}C ^{90}Sr $^{99}T_c$ ^{129}I $^{137}C_s$

^{232}Th ^{235}U ^{238}U ^{239}Pu ^{241}Am

What Do We Want To Know?

The pragmatic questions are:

- *How slowly does the radioactivity go into solution in the groundwater?*

- *What concentration does it attain?*
  Since the water movement will be very slow compared with the rate at which the wastes dissolve, we are concerned first and foremost with equilibrium solubility. Also, if only to relate behaviour on the geological time scale to that on the laboratory time scale, we will need to know about the mechanisms and kinetics of dissolution and leaching. The waste forms envisaged at present are glass blocks containing separated fission products and residual actinides fused into the glass and, alternatively, the uranium dioxide matrix of the used fuel containing unseparated fission products and plutonium. In the

same context, we may be interested in the rate of dissolution
of container materials.

- *How strongly is water transport of the radionuclides retarded
  by sorption on the backfill and sealing materials and the
  enclosing rocks?*
  Here again we are concerned with the equilibrium sorption
  characteristics for all available sorption processes. Also,
  we need to have sufficient knowledge of the sorption mechanisms and kinetics to relate geologic to laboratory conditions.
  Likewise we require a comparable understanding of the displacement and release processes which permit migration to proceed.

- *How do the pertinent properties of the waste form change
  with time?*
  Here we are concerned with changes in the solid structure of
  the waste form, such as devitrification and phase separation
  in waste-containing glasses. Thus we need to understand the
  mechanisms and kinetics of the solid-state transformations
  and the effect of these changes on the solubility of the
  wastes. Besides the transformations that might occur in
  the dry state, we also need to know what hydrothermal changes
  might occur.

It will be convenient to examine the underlying chemistry in
terms of interactions of pairs of constituents of the repository
and, finally, the waste-form stability itself:

> water-waste form (glass or UO_2)
> water-rock
> waste-rock
> waste-form - time (i.e. stability)

Water-Waste Interactions. It is appropriate to examine the
water-waste interactions first since this is an extension of our
previous interests in the high temperature solubility and mass
transfer of corrosion products in power plants (2) and our per-
ceptions in this area are therefore well developed.
    Solubilities can be obtained from free energy data, and vice
versa, by means of relations* such as the following (Equations 2,
3) which pertain to a simple oxide, A (= MO_x) giving solution
species B (Equation 1).

---

* a_B = activity of solution species B

  ΔG^o = standard free energy change of reaction 1

  C_s = saturation concentration of A in solution

  γ_B = activity coefficient of B

$$\alpha A + \beta H^+ + \tfrac{1}{2}\delta H_2 = \delta B + \epsilon H_2 0 \tag{1}$$

$$\log a_B = \frac{-\Delta G^O}{2.303 RT\delta} + \frac{\beta}{\delta} pH_T + \frac{\gamma}{2\delta} \log P_{H_2} \tag{2}$$

$$C_s = \sum_B (\alpha/\delta)(a_B/\gamma_B) \tag{3}$$

Detailed discussions are available elsewhere (3,4). Solubility
data obtained in this way for crystalline UO_2 are shown in Figure
1 which also indicates how the solubility depends on acidity,
redox-potential, and carbonate concentration (5).

To use these expressions, free energy values are required
for aqueous ions and other dissolved species at elevated temperature. In lieu of experimental data, which are sparse, the high
temperature free energy values are calculated from room temperature data. Three principal methods of calculation are in use.
In order of increasing refinement, these are as follows:

1. The relation (4) may be integrated using the room temperature
value of the specific heats of the ions in solution*.

$$\bar{G}^O_T - \bar{G}^O_{298} = -\bar{S}^O_{298}[T-298.15] - T\int_{298}^{T} \frac{\bar{C}^O_P}{T} dT + \int_{298}^{T} \bar{C}^O_P dT \tag{4}$$

2. An equivalent expression in terms of entropies may be used
along with entropy values determined by the empirical Criss-
Cobble principle (6).

3. The calculation may be based on a recently evolved theoretical
electrostatic model for ionic hydration (7).

Using methods 2. and 3., my colleague, P.R. Tremaine, has
extended the calculated solubility data for UO_2 (like that in
Figure 1) and for $(U,Pu)O_2$ up to 473 K (5).

To obtain and verify such data as these and facilitate the
extension to the broad range of complex mixed oxide systems of
interest, we require:

---

* $\bar{G}^O_T$ = the standard partial molal free energy of formation
of a species in solution at temperature T in the hypothetical standard state of unit activity and at a
pressure of 0.1 MPa, from the elements in their
standard state at 298K.

$\bar{S}^O_{298}$ = the standard partial molal entropy of a species in
solution at 298K ($\bar{S}^O$ (H+) $\equiv$ 0).

$\bar{C}^O_P$ = the standard partial molal heat capacity of the
species in solution.

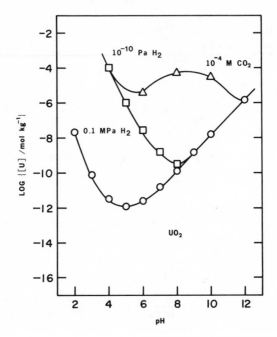

Figure 1. UO₂ solubility from thermochemical computations: concentration of U in a saturated aqueous solution at 25°C as a function of pH showing influence of redox potential (expressed in terms of eqilibrium hydrogen pressure) and dissolved CO₂. (○) 10⁵ Pa H₂; (□) 10⁻¹⁰ Pa H₂; (△) 10⁻⁴ M CO₂ and 10⁻¹⁰ Pa H₂.

- More and better solubility data for simple and complex oxides
  (glasses, UO_2 matrices) at elevated temperatures and pressures
  and improved techniques for measuring small solubilities.
  Solubility data for simple oxides provide thermochemical data,
  free energies in particular, which allow prediction of solubilities for complex oxides. Solubility data for some complex
  oxides are also necessary in order to verify the methods of
  predictions for complex oxides and establish key points.

- Heat capacity data for ions in aqueous solution over the
  temperature range 25-200°C. Such data for ionic species
  of uranium, plutonium, other actinides and various fission
  products such as cesium, strontium, iodine, technetium, and
  others are of foremost interest.

- Improved theoretical models for ionic hydration and the variations with temperature of the solvating properties of water so
  that the free energies can be more accurately extrapolated to
  elevated temperatures. These models must progress from simple
  monovalent ions to polyvalent and complex ions, e.g. Cs^+, Sr^{2+},
  Pu^{4+}, UO_2^{2+}, ions of Tc, I, etc.

- Complexation constants for important radionuclides, e.g. Pu,
  U, with common groundwater constituents, e.g. HCO_3^-, HSO_4^-, HS^-.
  Values are required for the 20-200°C temperature range or
  models which will permit extrapolation over this range.

     Water-Rock Interactions.   There are three main water-rock
interactions:

- The establishment of the groundwater composition by the rock
  and the backfill materials of the repository.

- Changes in surface structure of rock constituents by water
  action.

- Hydrothermal changes of rock composition.

     For slowly moving water deep in rock, the groundwater compo-
sition will be established by solubility equilibria. The theoretical knowledge and basic data requirements for these are
entirely analogous to those discussed above for dissolution of
the wastes.
     Electron microscope examination shows that freshly cleaved
rock surfaces undergo extensive modifications when they are
exposed to water. We need to know about such changes in order
to correctly relate laboratory results to long term behaviour.
     The direction of long-term changes in rock composition due
to the action of groundwater can be determined by thermochemical
calculations of the equilibria between various possible constituents (8). Appropriate thermochemical data are required. Exper-

imental studies of accelerated changes at higher temperatures
will provide kinetic information to estimate which changes may
proceed to a significant extent.

Waste-Rock Interactions. Under the heading of waste-rock
interactions, we have firstly the various processes which contribute to the sorption of radionuclides by the rock (Figure 2) and,
secondly, bulk changes due to hydrothermal interactions between
waste and rock leading to mass and activity transport.

Radionuclide migration will be controlled by surface adsorption and ion-exchange equilibria for which an important requirement is a knowledge of the free energies in solution of plutonium,
uranium and other actinide ions at elevated temperatures, as
already discussed in connection with solubility. Furthermore,
thermochemical data are required for the free energies of the
various adsorbed states on the surfaces of rock minerals. To
utilize these data, we require tractable theoretical and conceptual models for the chromatographic behaviour of solutions
advancing through cracked or fractured rock. Models have been
developed for alkali and alkaline earth ion migration in soils
(9) and simple feldspars (10). These models must be extended to
include solid-state diffusion, precipitation, and the decay and
growth of species through transmutation (11).

Thermochemical equilibrium calculations such as are used to
predict the direction of hydrothermal transformations of rock
minerals (8) can also be used to determine the direction of hydrothermal transformations of the waste form (12) and to estimate
mass and activity transport by water in the waste-rock system.
Since the time scale is short geologically, kinetic measurements
are essential. Because we are studying movement of radionuclides,
very slow changes can be observed by radioactivity counting techniques. It may also be advantageous to accelerate the processes
by employing elevated temperatures - say up to 300°C. Thus we
can expect development and application of very sensitive techniques for the study of hydrothermal interactions between solids
containing radioactive waste and the minerals of rocks. No doubt
these studies will aid progress in the understanding of geochemical processes generally.

Waste-Form Stability. If they occur at all, solid state
transformations in dry glass and UO_2 matrices will be too slow
under the temperature conditions of service to be observable in
the laboratory at the same temperature. Here we need to extra-
polate from high temperature laboratory conditions to low temperature service conditions. It will be desirable to develop an intimate knowledge of the processes of phase separation and devitrification of sodium borosilicate glasses at temperatures below the
softening point by meticulous application of electron microscopic,
X-ray crystallographic and other techniques. The glasses will
contain inactive elements representative of the fission product

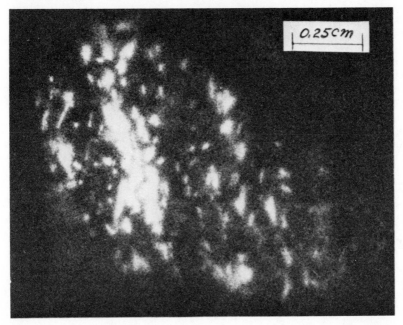

Figure 2.   Adsorption of radionuclides on rock minerals. (a) (Top) Photomacrograph of a polished granite surface after exposure to a solution of radio-cesium ($^{137}Cs^+$ + $^{134}Cs^+$). Dark areas are exposed mica crystallites. (b) (Bottom) Autoradiograph of same surface showing uptake of cesium by mica.

and actinide components of the wastes. Similar studies will be carried out for crystalline UO_2 matrices at temperatures above 1000°C. Phase separation may yield some phases which have higher solubility. Thus the effects of phase separation and devitrification on the solubility characteristics will be determined by measurements on glasses which have been deliberately transformed by appropriate heat treatment. These results will be used to test theoretical predictions using thermochemical calculations.

In the wetted condition, hydrothermal transformations of the waste form may occur at the temperature of service. Changes may occur either by dissolution and re-precipitation or by assimilation of water into the internal structures of the oxide matrices. The thermochemical essentials for hydrothermal recrystallisation have already been referred to previously in this paper. We also need to develop an adequate knowledge of the water-catalysed structural transformations.

Conclusions

Research and development is in progress for the final major stage in the large-scale exploitation of nuclear power, namely the underground disposal of radioactive wastes. Several areas have been identified where advances in basic chemical knowledge and data can help fulfil the practical requirements of this energy technology. In particular, we require more knowledge of the hydration, complexing and thermochemistry of actinide elements in aqueous solution at moderately elevated temperatures, their adsorption and ion-exchange properties on rock minerals and possible phase separations in glassy and crystalline matrices containing them. In answering these pragmatic questions, we shall undoubtedly stimulate advances in theoretical knowledge, particularly in connection with the properties of aqueous solutions and in geochemical processes. There will also be spin-offs in other technologies such as the location and exploitation of minerals and also in the disposal of noxious substances from other energy sources and industry generally.

Literature Cited

1. Inhaber, H., "Risk of Energy Production", Atomic Energy
    Control Board Report, AECB-1119, March 1978.
2. Tomlinson, M., "Dissolution: Questions for Energy Producers,
    Scientists and Teachers", Chemistry in Canada, (1977), 29(1),
    25.
3. Macdonald, D.D., Rummery, T.E. and Tomlinson, M., "Stability
    and Solubility of Metal Oxides in High Temperature Water",
    proceedings of the IAEA Symposium on the Thermodynamics of
    Nuclear Materials, Vol. II, p123, International Atomic Energy
    Agency IAEA-SM-190/19, Vienna (1975).
4. Macdonald, D.D., in Modern Aspects of Electrochemistry,
    J.O'M. Bockris and B.E. Conway, eds., Vol. II, Plenun Press,
    N.Y. (1976).
5. Tremaine, P.R., unpublished results.
6. Macdonald, D.D., Shierman, G.R., and Butler, P., "The Thermo-
    dynamics of Metal-Water Systems at Elevated Temperatures.
    Part I: The Water and Copper Water Systems", Atomic Energy of
    Canada Ltd., Report AECL-4136 (1972).
7. Tremaine, P.R. and Goldman, S., "Calculations of Gibbs Free
    Energies of Aqueous Electrolytes to 350°C from an Electro-
    static Model for Ion Hydration", Jr. Phys. Chem., in press.
8. Helgeson, H.C., "Evaluation of Irreversible Reactions in
    Geochemical Processes Involving Minerals and Aqueous
    Solutions - I: Thermodynamic Relations", Geochimica et
    Cosmochimica Acta, (1968), 32, 853.
9. Sung Ho Lai and Jwrinak, J.J., Water Resources Res. (1972),
    8, 99.
10. Fournier, R.O. and Truesdell, A.H., Geochim.Cosmochim.Acta,
    (1973), 37, 1255.
11. Burkholder, H.C., Cloninger, M.O., Baker, D.A. and Jansen, G.,
    Nucl.Technology (1976), 31, 202.
12. Shade, J.W., "An Approach to the Prediction of Waste Form
    Stability", Atlantic Richfield Hanford Company Report,
    ARH-ST-105 (August 1974).

RECEIVED September 25, 1978.

INDEX